Sustainable Urban Metabolism

Sustainable Urban Metabolism

Paulo Ferrão and John E. Fernández

The MIT Press
Cambridge, Massachusetts
London, England

MIT Press books may be purchased at special quantity discounts for business or sales promotional use. For information, please email special_sales@mitpress.mit.edu or write to Special Sales Department, The MIT Press, 55 Hayward Street, Cambridge, MA 02142.

This book was set in Sabon by Toppan Best-set Premedia Limited, Hong Kong. Printed and bound in the United States of America.

Library of Congress Cataloging-in-Publication Data
Ferrão, Paulo, 1962–
Sustainable urban metabolism / Paulo Ferrão and John E. Fernández.
   pages   cm
Includes bibliographical references and index.
ISBN 978-0-262-01936-1 (hardcover : alk. paper)
1. Sustainable urban development.   2. City planning.   3. Urban ecology (Sociology)
4. Globalization.   I. Fernández, John, 1963–   II. Title.
HT166.F378 2013
307.1'216—dc23
2012048911

10   9   8   7   6   5   4   3   2

Contents

Acknowledgments

We would like to thank the following individuals and organizations for their support in the writing of this book.

First, we would like to highlight the pivotal role of the MIT–Portugal Program in bringing the authors together to make this work possible. In particular, we would like to thank Dan Roos and Dave Marks for their undying support and wise council in this project. Without the MIT–Portugal Program, and others like it, our academic world would be less collaborative, cooperative, and effective across regional and international borders.

Second, we would like to acknowledge the support and inspiration from all of our colleagues in the industrial ecology community. The authors count this community as our intellectual home base, and we feel deeply privileged to be able to count on this substantial scientific community. The work on cities is just beginning and numerous colleagues from industrial ecology have been instrumental in guiding our own motivations and interests in the emerging field of urban metabolism.

In addition, both authors recognize the great privilege and responsibilities inherent in being members of our academic institutions; for Paulo, the Instituto Superior Técnico; for John, the Massachusetts Institute of Technology. Thank you to our home colleagues and staff in supporting and inspiring us.

John Fernández would like to thank Paulo for his gracious invitation and hosting of my research stay at the Instituto Superior Técnico in Lisbon, where the idea for this book was spawned and preliminary work drafted.

Finally, and most importantly, we would like to thank our families for their patience in accepting the absences and distractions that were an inevitable consequence of having this project on our minds for as long as it was.

This book is dedicated to them: Delmira, Sofia, Diogo, Malvina, Vita, and Lorenzo.

Introduction

We live in a world that is perceived to be rapidly changing, altering ways of living and disturbing a general sense of stability. This seems partly due to the increasingly complex interactions between economic, social, and environmental dynamics resulting from a cacophony of competing societal interests at the global scale. Many are considering the need to address these vertiginous changes in the way we are organized and interact with each other so that we may protect our wonderful, long-lasting, and unique habitat, our planet Earth.

If we try to identify the main drivers for changes having taken place during the last century, there seem to be two apparently contradictory trends: globalization and urbanization, each powered by frenetic technological development.

In fact, these two dominant trends are not contradictory, as globalization consists in establishing multiple linkages between countries through a network of communicating hubs of urban systems. In this network we find the mutually reinforcing relationship between globalization and urbanization. Globalization can thus be understood as a global network of urban systems supported by new infrastructures of energy, transportation, and communications developing at the pace of technological change, accompanied by trade liberalization and the growing importance of supranational rules, and fostering fierce competition among urban systems worldwide. Increasing globalization results in growing urbanization.

Globalization has been acting to influence, and been influenced by, urbanization, resulting in urban systems as engines of growth for the regional economies in which they are embedded. This is why global urbanization is proving irreversible and why most of the global population now lives in cities. This urban population now generates more than 80 percent of global gross domestic product and accounts for enormous flows of energy, materials, and water that serve urban residents with critical provisions while dispersing wastes with negative impact on the environment.

Urban centers also reflect the societal stress associated with rapid changes in the world economy. While cities concentrate most of the world's economic activities, including industrial production, producing enormous wealth, improvements in education, and access to markets, a significant proportion of the world's population with unmet needs continues to live in urban areas.

Urban centers concentrate large proportions of the middle- and upper-income population, converging the demand for goods and services at a single location and underpinning most of the resource demands from production worldwide. This demand will ultimately lead to the production of waste. In fact, as a result of urban centers concentrating demand for every manner of goods and services, and as they are certainly not sites for agricultural production or natural resources extraction, many of the environmental impacts generated outside urban centers can be directly linked to urban-based demands.

Thus, the quality of environmental management within urban centers, which include measures to increase resource-use efficiency and reduce waste generation, critically influences global sustainability.

Sustainability is therefore dependent on the way we collectively organize ourselves in growing urban centers. Doing so depends on the ways in which we conduct our analysis of urban systems, design and engineer them, and manage their multiple and complex interactions: economic, social, and environmental. Global sustainability is not dependent on the technological characteristics of global systems, but rather on the technology and design of local urban systems.

However, engineering systems are to be considered in a very broad sense if, as suggested here, we adopt the view of sustainability provided by John Ehrenfeld (2008), who argues that sustainability is a mere possibility that human and other life will flourish on the Earth forever. The notion of flourishing connotes not only mere survival but declaration of life as meaningful in terms of justice, freedom, and dignity. These are the attributes that an urban system should provide to their citizens, coupled with respect for and responsibility toward the environment.

This can only be achieved by promoting individual responsibility, such that every action taken by humans would entail a prior assessment of the potential harm of that action to the possibility of sustainability. That assessment needs to be informed by available data to be provided through a scientifically based framework of models and indicators.

As a consequence, in this book, we offer an intellectual framework to guide us through the challenges to come. Through industrial ecology as a metaphor that invokes nature as a source for new ideas to redesign systems for sustainability, we offer a roadmap of the diverse sciences, strategies, and tools that may help us in this journey. Finding the best ways in which to make good choices about our urban future suggested the title of this volume, *Sustainable Urban Metabolism*.

The Concept of Urban Metabolism

Urban systems are not a new phenomenon. Human beings have inexorably become urban creatures, as discussed by Baccini and Brunner (2012). From a metabolic perspective, this is clear. Our urban reality today is defined by high population density, high stocks of materials and exchange rates of information and goods, and a vital dependence on sources and disposal sites for energy and matter far beyond a settlement's borders. They point out that cities have declined concurrently with the goods they produce and market. Cities have become our most complex anthropogenic systems.

Each urban system has its own traits that reflect a cultural and socioeconomic context, as well as the infrastructures that, all together, determine the interaction of the urban system with nature, particularly how it exchanges matter and energy, establishing a characteristic "urban metabolism." In fact, the metabolisms of our urban systems have been increasing in volume, as the material turnover of a modern city is about one order of magnitude larger than that in an ancient city of the same size (Brunner, Daxbeck, and Baccini 1994).

Brunner (2007) shows that, in contrast to ancient cities, this contemporary metabolism consists mainly of water and air for human activities to "clean," "reside," and "transport" fuel. Construction materials are the second greatest flow, and food and other goods for consumption follow. Technologies developed in the nineteenth and twentieth centuries, such as fuel- and electricity-driven engines, electronic circuits, chemical synthesis of polymers, pharmaceuticals, and fertilizers formed the base for the tremendous economic success story of cities. During the next few decades, new megacities, particularly in Asia, will catch up with the material turnover of U.S. and European cities and may even surpass their wealth and grandeur.

Cities at present are mainly linear reactors: their metabolism consists of consuming materials and goods from elsewhere, transforming them in buildings, technical infrastructures for energy or water supply, communications or mobility, and wastes, which are summarily discarded, typically involving very limited reuse or recycling. During the long process of growing a city, the input of materials outweighs the output, resulting in growing stocks of increasingly valuable materials which can also be regarded as future resources, if we imagine adopting a different perspective.

Urban metabolism can be analyzed in terms of four fundamental flows or cycles—those of water, materials, energy, and nutrients, according to Kennedy et al. (2007). They suggest that differences in the cycles may be expected between cities due to age, stage of development (i.e., available technologies), and cultural factors. Other differences, particularly in energy flows, may be associated with climate or with urban population density.

It is thus suggested by Kennedy et al. (2007) that a full evaluation of urban sustainability requires a broad scope of analysis. Urban policy makers should be encouraged to understand the urban metabolism of their cities. It is practical for them to know if they are using water, energy, materials, and nutrients efficiently, and how this efficiency compares with that of other cities. They must evaluate to what extent their nearest resources are close to exhaustion and, if necessary, consider appropriate strategies to slow exploitation. It is apparent from Kennedy et al.'s review that metabolism data have been established for only a few cities worldwide, and there are issues of interpretation due to lack of common conventions. Clearly, there is much more work to be done. Resource accounting and management are typically undertaken at national levels, but such practices may arguably be too broad and miss elemental understanding of the urban driving processes.

In any case, the overriding purpose of all of this work is to contribute to the promotion of quality of life without compromising the capacity of Earth's finite resources—that is, sustainable development. This objective is particularly relevant in the urban context. Despite the fact that urban systems occupy only about 2 percent of the world's land, they are the hubs through which virtually all international commodities trade flows, which is to say where the resources flow and where a significant amount of waste is generated. It is thus evident that global sustainability depends on how we manage urban systems in the twenty-first century.

This grand challenge of promoting urban sustainability ultimately depends on attending to the details that are supported by science without losing our grasp of the big picture. This requires promoting the knowledge that supports development of a systems perspective, and embracing the multiple scientific domains involved in this complex perspective.

The practical implementation of these concepts requires smarter infrastructure that provides urban citizens with information and tools to translate technological development into new solutions for cities that are more resource-efficient and more livable, while also serving as engines of innovation and creativity.

The industrial ecology metaphor, as an intellectual framework, suggests that we ought to look to nature as a source for new ideas about how to redesign for sustainability. Ecosystems offer opportunities to learn from observing resilient, robust, long-lived ecological communities as examples of sustainable systems. For example, we can redefine our cities to adopt economies that incorporate a circular metabolism in which wastes are converted back into nutrients (raw materials) that can be reintroduced in the manufacturing process by detritivores, which in our world may be represented by recyclers.

Studying and quantifying metabolic flows provides the basis for discussions of the desirability of changes in the scale or type of a city's metabolism, particularly by correlating different economic activities with the materials flows they require and generate, and how such changes might best be accomplished through targeted policies. The metabolism approach provides measures of the resource intensity of urban systems and the degree of circularity of resource streams, and is helpful in benchmarking and identifying opportunities for improvement.

Understanding the urban metabolism of our cities and how we can modify urban respiration is not a simple task and, in its early stages, will require a holistic vision of the key scientific domains required for the analysis of the complex interaction of urban dynamics and the environment. Providing this systems approach constitutes the primary motivation for this book.

A Walk through the Book

In essence, this book is intended to provide a systems-oriented approach that establishes useful links between environmental, economic, social, technical, and infrastructural issues that lead to an integrated, information-intensive platform from which alternative urban technologies and ecologically informed planning may be achieved. Contrasting and opposing urban objectives can be considered holistically, avoiding the usual pitfall of solving problems in one sector only to create new ones in others.

We aim at contributing to the unification of currently fragmented analytical perspectives and methodologies toward the development of the innovative concept of urban metabolism and its integration of the technical, social, and economic activities that govern urban resource flows. The approach takes advantage of the recent revolution of communication-system infrastructures and combines the development of a new generation of IT systems, which network devices and manage the enormous amount of resulting data, with new scientific developments in specific areas such as "knowledge urbanism." Together these platforms will promote resource-efficient, knowledge-driven urban environments.

The book is organized into four sections. The first section is comprised of three chapters. The first presents industrial ecology as a primary perspective with which to consider sustainable development and urban futures. Chapter 2 outlines the various challenges that confront our urban future. The final chapter of this section illustrates the diversity of disciplines that have been contributing in some productive way to the consideration of an alternative and resource-efficient urban world.

The second section guides the reader through the tools and methods available to assess and understand the resource intensity of society generally and urban economies in particular. Section III begins with a sampling of urban sustainability approaches and measures of progress toward urban resource efficiency and ends with chapter 9, which presents the integrated approach of urban metabolism.

Section IV offers a synopsis of the shared and distinct challenges to be found in approaching urban sustainability in developed and developing economies.

These four sections are intended to present both challenges and opportunities that currently exist in the emerging field of urban metabolism.

I

Urban Metabolism: Defining a Field

1

Industrial Ecology: A Metaphor for Sustainable Development

Urban systems are increasingly becoming the locus of consumption and engines of economic growth in a globalized world, which everyday sees more people flowing from rural areas to cities. This movement contributes to separate humans from nature and eventually for us to lose the sense of the limits to growth, which is inherent to ecosystems, the basis for the natural world sustainability.

The global ecosystem relies mostly on solar energy inputs while operating with material cycles and energy cascades, and this constitutes the metaphor that provides the intellectual framework for industrial ecology. Industrial ecology combines a metaphor with a set of methods and models that help to quantify the interactions of urban systems and the environment, particularly the consumption of nonrenewable natural resources, frequently imported from beyond the immediate urban system boundaries, and the release of different types of wastes and emissions.

The industrial ecology framework is used throughout this chapter to provide an intellectual and analytical framework to aid citizens in becoming aware of the consequences of their behavior at different levels and scales, as well as to provide guidelines to inspire leaders of the future in promoting a paradigm shift that will transform urban systems into the lighthouses of sustainable development to which we have the right to aspire.

A Story That Cannot Be Repeated

Imagine a wealthy citizen driving alone in a major metropolitan city: the vehicle's air conditioning provides him a comfort the pedestrians who massively cross the roads under the burning sunshine cannot experience. He arrives at a shopping center where he joins a huge crowd to consume products that were produced thousands of kilometers away and fruits whose mother trees he does not know, but that will allow him to enjoy a light and healthy meal at home, comfortably looking at a television, on which he watches a documentary that discusses the unsustainability of the current world.

However, he believes science and technology will be able to solve "that problem," through a technological fix soon to be available and, as he might have spent the day alone, even if surrounded by people, he turns to the Internet as a forum to interact with a social network that the new information technologies have provided him, particularly in an urban environment.

At the end of the day, this typical citizen of a modern city may have contributed to the use of about 15 kilograms of fossil-fuel equivalents. This is about five times what more than 80 percent of people in the world consume in the same day, and if this usage were adopted by all of us, it would more than triple fossil fuel energy consumption worldwide—a level clearly unfeasible given the current global environmental pressures we are facing.

This is a vision of the world as a machine, which humans control from urban areas, where they concentrate and where they learn how to dominate the rest of the living and inanimate world through increasingly complex technology. This model is embedded in our culture, and indeed, a growing number of people look to cities as opportunities for a better life, a place in which we view ourselves as autonomous, isolated individuals driven only by self-interest and where nature is an endless source of resources, far away, that we fail to understand or to connect with.

In this scenario, we seek for efficiency to obtain more profit, to have more, to consume more, to waste more and, eventually, we confuse the "being" with the "having." Therefore, we spend most of our time in dissipating goods that others "not-so-wealthy" have produced at a lower cost and whose waste products at the end-of-life will be taken care of by others and by the "machine" we perceive as the world. Meanwhile, concerns over equity and justice are left to other and exogenous social mechanisms.

This artificial separation between the individual and nature is a sign of modernity and metropolitan life, and although very recent, it constitutes a major challenge for the sustainability of humankind. According to the prevailing view, one buys whatever he can afford to purchase, without any feedback on how this might be unfeasible from the point of view of the availability of natural resources or the capacity of nature to process our residues and pollutants. We have already gone too far beyond our limits in both dimensions.

Filling this gap between the urban dweller and nature will require change that, as discussed by John Ehrenfeld (2008), needs to go beyond the Cartesian model of objective (economic) rationality—that is, beyond an optimistic view about technology and technological change, or individualism and freedom of choice. It is clear that change needs to be built on a better understanding of the complex interactions that characterize urban systems and nature, which may provide the individual with informed options. In this context, we may, as suggested by Ehrenfeld (2008), need to begin to redesign parts of the structure that underlies the process of achieving sustainability, based on new ideas from every trace of sustainability we can locate. Nature itself can be a source of inspiration, which we will explore as we proceed in this volume, understanding that the last decades of our story cannot be repeated and that we need to find new pathways for the future.

This book is aimed at providing a better understanding of urban material flows and their interaction with nature, making use of the urban metabolism concept, which draws from an analogy with the metabolic processes of organisms. Urban metabolism models quantify the inputs, outputs, and storage of energy, water, nutrients, materials, and wastes, and may provide individuals with essential feedback on the impact of their choices. As a result, these models may contribute to changing behavior and diminishing our disconnect with nature.

On the other hand, we look to nature as a source for new ideas to redesign sustainability. Ecosystems offer possibilities to learn from observing resilient, robust, long-lived ecological communities as examples of sustainable systems. We look at industrial (or any anthropogenic) systems with an ecosystems perspective, as suggested by industrial ecology, which metaphorically blends ecological systems and industrial economies. The industrial side suggests that society can move toward more sustainable economies by embedding the principles learned from ecological systems into the design of companies and larger social institutions. For many, industrial ecology is paradigmatic in that it provides a new vocabulary for talking about and understanding sustainability. Here, we make use of industrial ecology as an intellectual framework to help spark a vision that might bring us to the realm of sustainability.

Paradoxically, one of the main barriers to looking at the future beyond the current cultural framework might be that the trend associating development with an exponential growth of materials consumption is extremely recent (only some decades in thousands of years of human evolution). Therefore, we might not yet understand it properly. In fact, the time scale of our actions in the world is unprecedentedly fast, much faster than the time scales nature can accommodate without major perturbations in ecological systems.

An Industrial Ecology Perspective on the Evolution of Urban Systems

A historical perspective on the interaction between man and the use of natural resources through the centuries might help identify major drivers for such recent and abrupt change. The first and obvious finding is that natural resources have always been abundantly used by society, which converts them to produce various commodities that provide well-being, according to the society's culture and economic development.

From this perspective, all societies have their own metabolism, in that they extract raw materials from nature, use natural goods, transform materials and energy within their economic systems, and provide goods for domestic and foreign demand. Part of the materials stay within the economy and result in the material stock of a society; this includes artifacts, machines, roads, buildings, and other infrastructure, and this has been true since the first humans lived on earth.

Human societies used hunting and gathering for subsistence for more than two million years. The first hunter-gatherers may have collected plants, fruits, and occasionally animals, predominantly carcasses of large animals killed by other predators or carcasses from animals that died by natural causes, and they were inherently nomads.

The transition into the subsequent Neolithic period is defined by the unprecedented development of agricultural practices. Agriculture originated and spread in several areas as early as about 10,000 years ago, which is to say recently, if we compare this to the million-years scale. Agricultural societies were fueled by sun and relied on photosynthesis for the conversion of solar energy in plants (biomass).

Hunter-gatherers and agrarian societies primarily differed in that the latter were typically sedentary and developed a set of skills to manage ecosystems in order to increase the output of useful biomass, which they used both as food and as energy for heating, cooking, and lighting in the form of fire.

The availability of biomass limited the transportation capacity associated with animal power at a terrestrial level, and only waterways allowed for long-distance transport of materials. This imposed limitations on the distances food and fuel could be transported, since the fuels were characterized by low energy densities. There were a few exceptions for the transportation of special goods, such as spices, whose economic value paid for them to be transported from Asia to Europe, first over land in small quantities, and later, after the fifteenth century, by sea, after Vasco da Gama, a Portuguese navigator, discovered the sea route to the Indian Ocean. It can be argued that these material flows (i.e., this trade) brought the world economy from the Middle Ages well into modern times and was the driver for European domination in the East, which has shaped the world as we know it today.

Despite this progress, the maximum amount of biomass that could be produced per unit of land limited the number of people sustained under the conditions of agrarian society, and consequently limited the dimension of urban areas, normally associated with trade centers.

The emergence of a new energy source characterized by much higher energy density, fossil fuels, together with the discovery of the steam engine in the late eighteenth century, gave rise to the Industrial Revolution, which broke through the limitations of an economy based on human labor and draft animals to move toward machine-based manufacturing and farming. For the first time in human history—and this was just two centuries ago—an increasingly small percentage of the population, with the support of machines and continuous scientific and technological development, could provide society with an increasingly complex diversity of products, and the disconnect between man and nature was born.

Throughout the nineteenth century and until the middle of the twentieth century, however, fossil fuel–powered urban-industrial centers coexisted with a rural matrix. The agrarian periphery perpetuated the conditions of the agrarian regime over many decades, as discussed by Krausmann et al. (2008). Meanwhile, all of the sociometabolic constraints stemming from the controlled solar energy system were abolished: energy turned from a scarce to an abundant resource, labor productivity in agriculture and industry could be increased by orders of magnitude, the energy cost of long-distance transport declined, and the number of people who could be nourished from one unit of land multiplied, allowing for an unprecedented growth of urban agglomeration. The transition from an agrarian to an industrial sociometabolic regime not only facilitated economic growth, structural change, and a certain worldwide uniformity in social forms and institutions, but it was inherently linked first to population growth and later to a surge in material and energy use per capita.

During the last fifty years of the twentieth century, economic growth and structural change were highly impaired across the world and varied significantly from country to country, but retained a very important characteristic: materials consumption was directly coupled with economic growth until a certain income threshold was reached, after which the aggregate material intensity of the economy seemed to stabilize. Canas, Ferrão, and Conceição (2003) show a sharp increase of materials consumption for economies that, over the years, increased their gross domestic product per capita between $3,500 and $10,000, and for which the materials consumption expressed in ton per capita also increased by a factor larger than three. This is characteristic of countries that might be considered as transitional economies, from underdeveloped to what we call developed countries, and is related to economic structural change, which might be associated with strengthening critical infrastructures (roads, railways, buildings, etc.), as discussed by Niza and Ferrão (2006).

This result is extremely relevant, as it shows that the pathway of rapid industrialization in a globalized world has become associated with energy- and material-intensive development. In this world, large infrastructures are being constructed and mobility is growing, via use of automobiles and air traffic, with bulk materials crossing the world in ships. Rising income has triggered the consumption of energy- and material-intensive goods and services.

In this context, if it is fair that developing countries should have the opportunity to catch up and provide their populations with a material standard of living others have enjoyed for decades, as is discussed by Krausmann et al. (2008), the preconditions are different than what they used to be in the first and second waves of industrialization. There is no wilderness frontier, no new worlds to be conquered anymore; fossil fuel energy supplies are running low and becoming more expensive, and the consequences of their use in the environment are of global dimensions. In most developing economies, a replication of the historically dirty material- and energy-intensive pathway could quickly lead to local and regional disaster. Politically, countries such as China have realized some of this and try to promote a "circular economy" approach to resource use. But to understand the full breadth of the problem, a sociometabolic perspective is required. The only adequate response to the dimension of the problem is a worldwide effort to invent, design, and experiment with a new paradigm that does not build human communication, creativity, and happiness upon gigatons and megajoules.

A Road Map to Urban Sustainability: The Contribution of Industrial Ecology

If we adopt the broad view of sustainability as provided by John Ehrenfeld (2008), who argues that sustainability is the mere possibility that human and other life will flourish on the Earth forever, where flourishing means not only survival but the realization of whatever we humans declare makes life meaningful—justice, freedom, and dignity—then these are the attributes that an urban system should provide to its citizens, together with respect for the environment.

In this context, sustainability is definitely not a technological characteristic of the global system, as the term sustainable development would imply. Yet its realization depends on the nature of the system. Responsibility, meaning that every action humans take should entail a prior assessment of its potential harm, is key to sustainability, and this depends on an assessment informed by available data provided through a scientifically based framework of models and indicators.

On the other hand, we have suggested before that we ought to look to nature as a source for new ideas to redesign sustainability, as ecosystems offer possibilities to learn from observing resilient, robust, long-lived ecological communities as examples of sustainable systems. The metaphor offered by industrial ecology provides a vision for a paradigm shift that may bring our society closer to sustainability. In this, industrial ecology offers two main contributions:

• an intellectual framework to support a paradigm shift, based on its metaphor, which can help identify new and innovative solutions to redesigning our development strategies.

• a toolbox with different methods and models to provide metrics and enable the establishment of a multidisciplinary framework within different fields, allowing the individual to perceive the effect of her actions and providing her with the necessary feedback (responses), namely through the DPSIR framework (i.e., drivers, pressures, state, impact, and response; see discussion later in this chapter).

Industrial Ecology as an Intellectual Framework to Support a Paradigm Shift
The power of the concept of industrial ecology, according to John Ehrenfeld (2000), lies in its normative context and in its potential to shape paradigmatic thinking. It is normative in the sense that the three features of the ecological metaphor—community, connectedness, and cooperation—are characteristics we should strive for in designing our worlds. When we look forward to becoming more like an ecological community, connectedness runs counter to the positivist, reductionist notions of knowledge and to the central idea of individual libertarianism—the realization of the autonomous self. Community becomes eroded as a result of the latter aspect. Competition is the main characteristic of modern corporate liberalism, and although it does not prevent cooperation, it is a limiting factor. It is not inherently normatively bad; it exists in natural systems, but in a balance with cooperation.

The notion of "limited competition," according to which individuals compete for a limited food supply, but not beyond the point that the biological success of their community becomes threatened, is key for sustainability and bridges the logic of sufficiency, as discussed by Princen (2005). He claims that prevailing principles of social organization—efficiency, cooperation, equity, sovereignty—are not up to the task of promoting sustainability. They may have worked in times of resource abundance, in an ecologically "empty world," a world where human impact was minor and where there was always another frontier. But they will not work now.

The age of efficiency in the early twentieth century was guided by "Taylorism," named after Frederick Winslow Taylor. His method made use of time and motion studies in different industries, and promoted the disaggregation of tasks to optimize rates of production. He introduced a system of specialization, separating skills and knowledge and rewarding workers for speed and output.

If it is true that efficiency made each product available at lower cost and with minimum resource consumption, this ubiquitous slogan of global industry has a dark side: the so-called rebound effect, in which efficiency produces more consumer surplus, which then creates more consumption, which may counterbalance and even overcome the potential benefits of efficiency in terms of the global use of resources.

The consumption society as described above lacks a principle that is offered by the industrial ecology metaphor, which is the principle of sufficiency. This principle includes concepts such as restraint, respite, precaution, polluter pays, and closing material cycles.

At this stage it is important to recognize, as discussed by Princen (2005), that humans do not always want more. Goods may be good, but more goods may not be better. In the more advanced societies we are already experiencing a movement of some highly educated professionals away from unpleasant, meaningless, unrewarding yet monetarily compensated work toward satisfying work (e.g., self-employment through consulting, producing niche-market products and services, or farming) that allows a broader and richer use of time and resources. In this sense, restraint is not self-abnegation, it is choosing less material use than what is possible in exchange for nonmaterial benefits; it is about privileging "being" instead of "having."

A new society focused on a new set of nonmaterial values can be designed if new customized services and products are as supported as massive production has been in the past. Eventually, this could be a relevant strategy to support the creation of high-added-value, culturally embedded local jobs, which are not so vulnerable to being transferred to the "low-wage" economy, country, or region of the moment that "efficiently" produces the commodities purchased around the globe.

In addition to the concept of sufficiency, which is at the level of human behavior, if we consider that at the ecosystems level, materials are circulated through a web of interconnections, with scavengers located at the bottom of the food web turning waste into food, the industrial ecology metaphor supports the concept of cooperation as an emphasis to establish a chain of industries that integrate the residues and end-of-life products of upstream industries in downstream manufacturing processes, thus avoiding the use of virgin raw materials and contributing to close the loop of materials use as do scavengers in an ecosystem.

Current policies are already addressing this idea. Although scavengers have historically lived in a social niche outside the mainstream of society, today, with the establishment of policy principles such as "extended product responsibility"—which determines that manufacturers are responsible for their products' entire life cycle, including end-of-life and recycling—recyclers are being integrated and upgraded in their business strategies.

Urban systems can also learn significantly from the use of the industrial ecology metaphor. While the power of a metaphor is to provide guidelines in moments of choice and decision, particularly those that we cannot anticipate now, table 1.1 provides some examples of the use of the metaphor in an urban context.

Industrial Ecology Toolbox: A Framework for Providing the Metrics of Sustainability
Each one of us makes quotidian decisions on the premise that they will not endanger our society, but particularly in an artificial, human-made context, such as an urban system, if prices do not reflect the evidence of environmental problems, then economic development will not decouple from resource extraction and waste generation. Therefore, we may promote intergenerational scarcity by exceeding the ecosystem's capacity to renew or absorb the materials we use or pollutants we generate. The only way to address these issues is by changing behavior, either through our own convictions or, more effectively, by promoting new policies, supported with

Table 1.1
Industrial ecology metaphor use in the design of sustainable urban systems

Topic	Human-developed systems	Natural systems	Lessons to sustainable systems
Energy	Central production, distribution networks	Distributed production in each plant by photosynthesis	Distributed production, e.g. building integrated photovoltaics
	Direct-use energy sources, for a single purpose	Cascaded use of energy	Integrated energy systems, e.g. cogeneration Energy-efficiency measures
Economy	Driven by money	Driven by solar energy, the basis for the food chain; focused on resources	The need to address new values, such as solidarity or voluntarism
	Focused on efficiency Focused on consumption	Focused on sufficiency	Need to internalize the limits of growth
	Focused on products	Dependent on services from other species, interdependence	Develop new business strategies focused on providing services rather than products
	Linear systems characterized by extraction of raw materials, product manufacturing, and waste	Circular systems	Promote recycling and reuse
	Material flows	Cascaded use of materials	Develop recycling infrastructures, or "scavengers"
Environment	Energy flows	Transformation of materials in energy	Energy valorization of wastes, whenever a better use is not available
	Tendency to promote monocultures	Diversity	Small-scale agriculture urban farming; context-integrated, added-value industries

Table 1.1
(Continued)

Topic	Human-developed systems	Natural systems	Lessons to sustainable systems
Transportation	Fossil-fuel dependent	Based on solar driven energy, through the food chain	Use of renewable energies for transportation
	Use of fuel-powered machines	Natural capacity of the individuals	Facilitate the use of human-powered transportation
	Vehicles are fuel-dependent and only allocated to transportation functions		Integrate electric vehicles in the network as storage facilities to enable the use of renewable energy
Society, the behavior of individuals	Focused on individual growth, efficiency	Focused on sufficiency and long-term flourishing of individuals	Promote solidarity, ethics, community services, and infrastructures
Infrastructure	Tendency to promote monoculture developments, namely in specific neighborhoods	Diversification	Adopt a multicultural strategy in urban planning

facts and data that may quantify the dimensions of the problems we are facing and identify possible responses.

The objective of the industrial ecology toolbox is to provide a set of scientific methods and models that may provide a framework of indicators to use as tools and explanatory constructs that permit policy makers and individuals to proceed without serious, frequent failures.

The industrial toolbox is introduced in the context of the length scales and time scales required to understand the metabolism of urban systems. These range from micro to macro in different dimensions, such as the economic, environmental, and geographical. Relevant macro tools are ecological footprint calculation, material flow analysis, and life-cycle assessment, which are characterized as follows.

- *Ecological footprint* This measure uses yields of primary products (from cropland, forest, grazing land, and fisheries) to calculate the area necessary to support a given activity.

- *Material flow analysis* This methodology is used to characterize the material flows that enter, accumulate, and leave a given economy during a period of time.

• *Life-cycle assessment*   A methodology for gathering information on the environmental impact of a product or service over its entire life cycle.

From Industrial Ecology to Urban Metabolism

The concept of urban metabolism derives directly from the application of the industrial ecology metaphor to urban systems. In this context, urban systems are open systems characterized by several interlinked subsystems—social, economic, institutional—that interact with the environment by consuming materials and energy they accumulate in the built environment and other stocks, and by rejecting different solid wastes and emissions to air and water, which are to be absorbed and regenerated by the environment.

The city can thus be viewed as an organism with a metabolism that can be studied. If we examine a city's metabolic flows, we can identify raw materials for construction, products, nutrients, energy, residues, and emissions, all with potential environmental impacts that extend well beyond the city limits.

The main objective of studying and quantifying a city's metabolism is to provide the basis for discussions of the desirability of changes in its scale or type, particularly by correlating different economic activities with the material flows they require and generate, and how such changes might best be accomplished via adequate policies.

First applied by Wolman (1965) to a hypothetical American city of 1 million people, there have been several metabolism studies of actual cities worldwide, including Lisbon, London, Sydney, Toronto, and Vienna.

Urban metabolism is influenced by several factors. The first is affluence, which determines level of economic activity. Climate and access to transportation infrastructure are other important factors in determining energy consumption and the flow of goods and trade. Urban form, including density, morphology, and transportation characteristics, with emphasis on public or private transport, influences both energy and material flows. Low-density, sprawling cities have generally higher per capita transportation energy requirements than compact cities. Technological development, use of vegetation, building codes, the costs of energy, or policies and infrastructure for residues management do also influence energy and material flows across the urban system.

Understanding the urban metabolism can be very helpful in providing suitable measures of resource exploitation and waste generation to be used as sustainability indicators. The metabolism also provides measures of resource intensity of the economy and the degree of circularity of resource streams, and may be helpful in identifying opportunities for improvement.

Characterizing urban metabolism is particularly relevant in helping urban policy makers improve the metabolism of their cities, as well as understand the feedback mechanisms that result from human pressure on urban areas. The DPSIR framework discussed in the following section is one such tool.

An Integrated, DPSIR Approach to Urban Metabolism

Alberti et al. (2003) have analyzed the opportunities and challenges for studying urban ecosystems and concluded that current studies of urban ecosystems use such simplified representations of human-ecological interactions that their system dynamics cannot be fully appreciated and understood. For example, most ecological studies treat urban areas as homogeneous phenomena and combine all anthropogenic factors into one aggregated variable (e.g., pollution load, population density, total paved area); thus, they represent urbanization as unidimensional, which is unrealistic, as urbanization is multidimensional and highly variable across time and space. Socioeconomic studies, on the other hand, highly simplify and rarely discriminate among different and complex ecological and biophysical processes.

To fully integrate humans into ecosystem science, Alberti et al. (2003) proposed a conceptual model that links human and biophysical drivers, patterns, processes, and effects (DPPE), which we will call a DPPE model. In this conceptual model, *drivers* are human and biophysical forces that produce change in human and biophysical patterns and processes. *Patterns* are spatial and temporal distributions of human or biophysical variables. *Processes* are the mechanisms by which human and biophysical variables interact and affect ecological conditions. *Effects* are the changes in human and ecological conditions that result from such interactions. For example, population growth in an area (driver) leads to increased pavement and buildings (patterns), leading to increased runoff and erosion (processes), causing lower water quality and decreased fish habitat (effects), which may lead to a new policy to regulate land use (driver).

This model could represent the interactions between human and biophysical patterns and processes, as well as the feedbacks from these interactions. In the Alberti et al. (2003) conceptual model, both biophysical and human agents drive the urban socioeconomic and biophysical patterns and processes that control ecosystem functions. This framework would be able to explain how patterns of human and ecological responses emerge from the interactions between human and biophysical processes and how these patterns affect ecological resilience in urban ecosystems. The model could therefore help test formal hypotheses about how human and ecological processes interact over time and space. It could also help establish (a) what forces drive patterns of urban development, (b) what the emerging patterns are for natural and developed land, (c) how these patterns influence ecosystem function and human behavior, and (d) how ecosystem and human processes operate as feedback mechanisms. Without a fully integrated framework, scholars can neither test hypotheses about system dynamics nor produce reliable predictions of ecosystem change under different human and ecological disturbance scenarios. Such knowledge is critical if managers and policymakers are to control and minimize the effects of human activities on ecosystems.

Here, we extend this DPPE model to a driver-pressure-state-impact-response (DPSIR) model, which provides an integrated approach and framework of indicators useful to urban metabolism. The major indicators required to promote an integrated analysis and the fields of knowledge involved in their quantification will be discussed throughout the book. This section will simply outline the main features of this conceptual model.

A conceptual model applied to urban metabolism is primarily intended to provide a framework to represent interactions between humans, urban systems, and the environment, as well as the feedbacks from these interactions. It is a highly aggregated model providing a general framework for system analysis, derived from more general considerations of the interactions between human systems (anthroposystems) and ecosystems, the city being a particular case of anthroposystem, as suggested by the urban metabolism approach.

A DPSIR conceptual model is based on indicators that result from a variety of multidisciplinary methods and models and that are particularly significant in the understanding of a phenomenon (for example, activity of a given economic sector as a proxy for material consumption). These indicators are thus used in the assessment of urban activities (transportation, industry, tourism, etc.). They constitute a flexible method to study complex phenomena, like the land use–transportation–environment interaction. They allow for a descriptive approach to be used even in the absence of a comprehensive theory of the phenomena to be analyzed. They can integrate both quantitative and qualitative pieces of information, and they answer the practical needs of public decision makers.

The use of a DPSIR model contributes to render explicit causality effects between different actors and consequences and ensure exhaustive coverage with respect to the phenomena retained in the model.

The DPSIR models are an extension of the pressure-status-response (PSR) framework developed by the European Organization for Economic Cooperation and Development (1993) as a common framework for environmental evaluation. Environmental problems and issues were taken as variables to show the cause-and-effect relationships between human activities that exert pressure (P) on the environment, the resulting changes in the state of the environment (S), and the responses to the change in the environment (R).

This PSR model was further enhanced by the European Environmental Agency (1999) to become the driving force–pressure–state–impact–response (DPSIR) framework, in order to provide a more comprehensive approach to analyzing environmental problems. This systems perspective suggests that economic and social development, which are common driving forces (D), exert pressure (P) on the environment and, as a result, the state (S) of the environment changes. These changes then have impacts (I) on ecosystems, human health, and other factors. Due to these impacts, society responds (R) to the driving forces, or directly to the pressure, state, or impacts through preventive, adaptive, or curative solutions.

The practical application of the DPSIR model is substantially based on success-fully quantifying each D, P, S, I, and R element and then specifying the various relationships between them. This involves the challenge of assembling a great deal of information that can serve to populate each of the elements of the DPSIR frame-work (measures or metrics) in ways that allow for the development of informative and appropriate assessments (indicators) that reflect the causal relationships between human activities, environmental consequences, and responses to environmental changes. The task of defining measures for each of the elements of the DPSIR framework (drivers, pressures, state, impact, and response), finding good data to quantify each, and formulating the most effective DPSIR indicators has been ongoing since the formulation of the framework itself. Of course, the ultimate success of the DPSIR framework in an urban context is associated with the ability to act upon feedback loops that inform policy makers to make better policy. Later in this chapter, it will become clear that doing so under the political reality of many cities and their regional and national governments is an enduring challenge, possibly more daunting than any other.

DPSIR in the Urban Context

Indicators are to be organized according to their effectiveness in relating the ele-ments of the DPSIR framework to one another, and sometimes in identifying the actors that are involved in the particular phenomenon the indicator is tracking. First, it will be helpful to describe metrics used to quantify each of the elements of the DPSIR framework and indicators that relate them to one another, and then discuss the identification of actors that can influence the intensity and direction of the dynamics of this framework.

Typical measures of driving forces include economic, social, and demographic attributes of society such as population, age composition, and household size. In an urban context, this involves delineating an urban population and characterizing its behavior.

Drivers lead to pressures. The environmental pressures resulting from human activities (emissions, resource use, and land use) are a function of two types of vari-able: the level of these activities, and the technology applied in these activities. For example, the emissions associated with a given process are the product of the level of activity and an emission factor, which reflects the technology of the process under scrutiny. The modeling of pressures therefore requires methods that take into account these two types of variables.

The technology variables can be reflected by emissions factors, resource-use factors, or land-use factors. In some cases these variables can be calculated from basic process information, but in other cases field information will be required. The variables accounting for level of activity are of an economic nature, since they reflect the behavior of citizens (consumption) and levels of production.

Responses that might influence the interaction between D and P include, among others, the eco-efficiency of various technologies used by consumers, changes in production and consumption patterns, and the lifestyles of the population. Included here are factors that are sometimes difficult to measure, such as the willingness of a population to adopt and use renewable energy and other resource-efficiency practices.

In the urban context, this is an extremely important aspect of the DPSIR dynamic because it might be argued that willingness to adopt eco-efficiency measures is strongly influenced by "word of mouth" and education programs. It is also suspected that the spatial compression of urban spaces compels the population to more readily consider measures that improve local environmental quality and possibly even resource efficiency. For example, it is more likely for urban residents to use mass transportation in a spatially compressed urban region in which traffic congestion results in delays and inconvenience.

It is clear that D will continue to increase in many of the world's cities. That is, urban population is increasing, both in absolute numbers and relative to rural populations. One grand challenge will be to reduce global P under the likely prospect of increasing urban D.

Common measures of pressure, **P**, include phenomena that compromise primary net production, degrade renewable material and energy sources, exhaust nonrenewable material reserves, endanger biogeochemical cycles and biodiversity, and otherwise prevent the environment and resident ecosystems from maintaining their integrity and flourishing. Of course, as related above, these pressures are results of drivers of socioeconomic and sociocultural origin, but can be measured independently of these origins. Thus, unsustainable logging of endangered hardwoods, depletion of fossil water reserves, accelerating mineral extraction, fragmentation of wildlife habitat, and many other pressures can each be quantified in their own way.

In the urban context, pressures can be very difficult to fully account for because of the distance at which a pressure is driven by urban socioeconomic activities. That is, pressure may be applied on a forest, body of water, mineral source, or other extraction point, at a great distance and sometimes halfway across the globe from the urban activity that acts to drive that pressure. Today this propagation of pressure at great distances is facilitated by international trade and interdependent global financial markets. In any case, it is possible to track the relationship between urban drivers and associated pressures and have some effects on global P. One growing movement involves local food production. Limiting the international movement of biomass, especially that serving cities, is one pathway toward reducing global P even under an increasing urban D.

Efforts to relate P to S include, for example, attempts to quantify the rate at which the pressure is acting; the rate of logging in a certain forest providing construction materials to the city, the rate of agricultural to urban land conversion to serve perimeter developments of low density, or the rate of water extraction to serve a particular city. While the general relationship between P and the state of the environment, S, is often clear, particular causal links may not be. As a result, the relationship between these pressures and the state of the environment may then be described in two different ways:

1. empirical models constructed from monitoring data related to pressures and changes in state, and;

2. analytical models constructed on the basis of physical, chemical, and ecological process modeling.

Monitoring data is used to calibrate these models. Analytical models may be more satisfactory from a scientific point of view, but are often very complex to develop and require enormous amounts of input information. A compromise is the use of "meta-models," in which the analytical model is used to calculate the cause-effect relation under certain specified conditions, and from the calculated results an empirical relation is constructed between pressure and state.

Indicators of the state of the environment include quantities that describe the condition of the physical environment, biological components of the environment (flora and fauna), and chemical concentration within a certain area. These measures include actual physical elements of the environment (for example, amount of wetland and water quality) that then serve to inform indicators that are effective in assessing the resilience, diversity, and capacity of the ecosystem services attributable to that environment.

Indicators that relate S to I outline the ways in which the environment and ecosystem services are dependent on a complex web of threshold conditions served by appropriate inflows and outflows. Some of these thresholds are governed by smooth gradients of water quality, flora biomass density, or other measures of environmental health. Others are governed by dynamics that indicate a step function between environmental health and collapse. Of particular importance here is the carrying capacity of a particular aspect of the environment. Exceeding that capacity may be a consequence of a particularly intense impact.

Impact indicators assess the social and economic functions of the environment, such as provision of adequate conditions for health, resources availability, and biodiversity. In an urban context, impact can often be found in health-related problems, such as the incidence of asthma from the presence of high levels of smog-related particulates from transportation or hospitalizations and even death from heat stroke due to the prevalence of the heat island effect.

Response strategies return us to the beginning of this cycle: drivers. Response measures refer to responses by different groups in society, as well as government initiatives to prevent the negative consequences to the environment, improve the conditions of the environment, or to adapt to changes in the state of the environment. While R is linked directly with P, S and I, it is the link with D that underlies, to a great extent the logic of DPSIR.

Therefore, indicators that link response and drivers are a critical element of the utility of the DPSIR framework. Yet it is precisely at this linkage point that the realities of municipal governance and economic development often compromise a truly effective feedback mechanism. For example, policies to promote urban green building practices that would lead to greater energy efficiency are notoriously lacking in post-occupancy monitoring of energy use. While it may seem obvious that building according to energy-efficient policies would lead to reduced energy use, this may not be the case in many situations if the existing policy incentives do not induce the use of effective technology solutions, such as energy-efficient lighting. In addition, there is some evidence to suggest that improved building systems, even if better prepared for energy efficiency, are leading to higher energy consumption in the form of more air conditioning to serve a more demanding comfort expectation. This is the danger of the emergence of the rebound effect in the absence of information that leads to an informed feedback loop.

As a consequence, the value to government in proposing and implementing green urban policies cannot be limited to their adoption as policy, but must necessarily be extended to monitor their effectiveness in actually improving resource efficiency. For the DPSIR model to work, investment in assessing the response is critical in ensuring that a policy has more than expedient political value.

In assessing the sustainability of urban activities, these interactions are fundamental to understanding the system dynamics. This motivated the establishment of a more coherent theoretical framework for the assessment of urban sustainability and to use it as a first step to derive a more specific framework for urban metabolism.

Our conceptual framework combines the causal approach of the DPSIR model with the main assumption of the extended urban metabolism model: that is, that the city is an anthropogenic system whose main objective is the satisfaction of human needs. This framework is proposed in figure 1.1, and considers two main classes of impact:

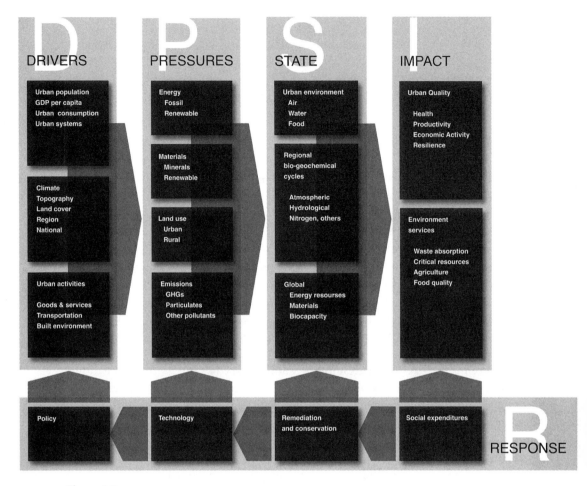

Figure 1.1
A DPSIR framework for urban metabolism.

- Urban quality: living conditions for city inhabitants.
- Environmental services: biogeochemical systems and processes within and beyond the urban system and the effect of their condition on humans and other species.

Human needs are the cause of urban activities and the basis for drivers. These then cause, on the one hand, the consumption of resources and the emission of waste and pollutants but, on the other, the satisfaction of human needs.

Emissions to the environment and resource consumption determine changes in the quality of the urban environment (they also contribute to changes in the quality of the regional and the global environments). However, the quality of the urban environment also contributes to the satisfaction of the needs of the inhabitants of the city.

These impacts are caused by the aggregate and often cumulative effect of various emissions, land use conversions, mobility, energy, and materials consumption in the service of the fundamental urban activities required by inhabitants.

It is clearly a much-aggregated model: every element and every link between elements represents a complex process. It is also a causal framework. The links between elements represent cause-effect relationships, which are modeled by multidisciplinary approaches that combine expertise in different fields. Arriving at robust and practical linkages between these various elements requires addressing them from multiple perspectives, including sociological, economic, environmental, and ecological. It also requires the involvement of multiple disciplines and professions, including engineers of many kinds, architects and designers, political scientists and policy experts, and members of the business and finance communities. The opportunities and challenges of this highly multidisciplinary effort are addressed in chapter 5.

In this context, the response phase of the DPSIR model entails a political dimension that, in an effective democratic regime, requires a wide spectrum of specialized knowledge and a well-structured framework to provide data adequate to inform the decision making of different actors. In fact, there is an indisputable link between information, comprehensibility, and democracy. The democratic form of government rests on two assumptions: that the average citizen can understand the system in which he is embedded, and that he can foresee the consequences of his actions sufficiently well enough to take responsibility for them. This is particularly difficult when we face the complexity of urban systems, which are characterized by their growing number of overlapping subsystems and increasing variations in interrelations, but addressing it properly is the key for urban sustainability.

In addition, an analysis of our highly technological, specialist-dependent, and mass media–influenced society leads to deep-seated doubts, summarized by von Wright (1989): "It is possible that the complications of the industrialized and technified society are so great that democratic participation in the public decision processes in the long run must degenerate to an empty formality of either assent to, or protest against, incomprehensible alternatives."

In conclusion, it is clear that urban systems are complex. Their sustainability requires a new form of knowledge, based on new visions and monitoring models such as those offered by industrial ecology and the DPSIR, which are not only technical but combine the technical and social. We are speaking of the need for a new science paradigm capable of dealing with complex sociotechnical systems, in which the system-builder who finds the key to sustainability becomes the hero of the history.

2

Urban Metabolism: Resource Consumption of Cities

Urbanization has been a driving force throughout human history, culminating today in what many believe will be a historic urban twenty-first century. A timeless consequence of the steadily increasing urbanization of civilization has been the associated erosion of our natural capital and the now global draw on every kind of useful resource. Studies now indicate that the past century was extraordinarily proficient in extracting resources that have fueled the transition from a world consuming renewable materials to one that requires nonrenewable metals and minerals to build up our urban infrastructure and building stock. The rapid urbanization of enormous numbers of the world's population has placed the resource intensity of our cities front and center in discussions regarding critical resources and global climate change. This chapter will review these themes and present a framework for considering the links between the urban world, urban activities, biogeochemical cycles, and the material and energy flows.

Urban Consumption

Imagine an architect, working on an upper floor of a skyscraper in New York City overlooking Wall Street. The year is 1996 and the global real estate market is once again expanding; a welcome contrast to the lean years earlier in the decade. Even the U.S. economy is gathering some hard-fought momentum, and once again, skyscrapers are rising in Manhattan. The architect is employed by a large, well-established international design and engineering firm specializing in large-scale projects, especially buildings serving the global commercial real estate market. This architect regularly selects materials for buildings that will eventually house financial services companies, insurance firms, law practices, and even large architecture businesses similar to his own. Much of the output of the design firm has moved into international markets: Malaysia, Japan, China. This designer travels frequently to Asian cities that are rapidly expanding their commercial office building stock; Shanghai, Jakarta, Kuala Lumpur, and lately, Beijing China. Lately however, the focus has turned back to domestic real estate development and even construction in his own backyard, New York City.

The architect is finalizing the design of the facade of a class A[1] commercial office building in midtown Manhattan. It is a relief to be working on a local project, though the designer must privately admit that the building design is not strongly influenced by the city in which it will be built. The form of the building, its organization and systems and the materials selected could easily have been adapted to, or from any one of many recent Asian design projects—and in fact, the main elements of building designs from the firm tend to migrate from continent to continent, searching for an owner willing to pay for construction.

In any case, the materials for the glass, aluminum, and stone silicone-sealed curtain wall now need more specificity. Low-emissivity glass, a relatively new product, is being considered, as is a whole range of anodizing colors for the aluminum mullions. The stone has yet to be specified, so a colleague down the hall is consulted and a range of stones are shortlisted.

The stone that is finally chosen is granite from the quarries in Minas Gerais in the southeast of Brazil. The selection of the stone will require many thousands of kilometers of travel, not only for the architect and a colleague or two but also for several tons of the stone itself. First, the architect and an assistant will travel to Brazil, where they will meet with the quarry representative and stone seller and then visit the quarry itself to inspect samples of the stone that is under consideration. The final selection will need to be made in the field, with the quarry providing several representative samples of the candidate stones, cut to thin slices, for the architect to take away and use for quality control when the final stone slabs are eventually delivered. Then a trip to the manufacturing facility located in Italy, where the stone will be shipped, is necessary to inspect the slabs as they are liberated from the boulders. In 1996, Italy dominates the global dimension stone industry by processing the material more reliably and less expensively than anywhere else in the world. It is common for stone boulders and slabs from every corner of the globe to pass through the Italian stone manufacturers of Carrara, located in the north, 100 kilometers west-northwest of the city of Florence. In 1996, the year of this transaction, Italy provided the United States with 45 percent of its dimension stone imports, amounting to roughly 25,000 metric tons out of a total of 55,100 metric tons (United States Geological Survey 1996; Taylor 1996). Total apparent consumption of dimension stone in the United States in 1996 amounted to 1.37 million metric tons, while in 2010 it hit 14 million metric tons (United States Geological Survey 2011; Dolley 2010). In 2010, the total value of imported dimension stone reached over $2.4 billion, and Brazil was the leading exporter, providing 44% of the material by value. International transactions in dimension stone have undergone dramatic change since 1996. China is now a major supplier of granite to the United States, providing 22% by value while the value of Italy's stone imports are down to 13% (Dolley 2010).

Once the stone is finally delivered to the site in midtown Manhattan, the architect inspects the pallets heavy with slabs and may or may not be satisfied with the general quality. If not, the representative stone samples in the architect's possession are invoked to argue for sending the stone back and replacing them with better slabs; more international travel for the rock. However, if it is approved, the stone is unpacked and prepared for erection onto an architectural curtain wall. Of course, final selection of the pieces that will be used in the wall will require laying out sections of stone and inspecting each one for color consistency and lack of any occlusions or other visible blemishes. Any number of slabs may still be sent packing. In a couple of weeks this process is well underway, and the architect is satisfied with the selection of Brazilian stone and its qualities.

It just so happens that the firm has just received word of a rapidly expanding building market opening in the Persian Gulf. Oil was discovered in the United Arab Emirates 30 years before, in 1966, and the urbanization of Dubai is accelerating beyond all expectation. Despite the flight of investment capital and trade volume from the city during the 1990 Gulf War, it is now clear that Dubai is in the midst of a vast expansion buoyed by oil revenue and real estate speculation. Rumors of an unprecedented building boom that would last well into the new millennium are circulating to every large New York City firm. This stone might just be the perfect material for the firm's first project there.

This tale is cautionary only in the sense that it documents the ease with which one individual, a well-educated modern professional, can set forth the movement of massive amounts of materials and the expenditure of enormous amounts of energy globally in the effort to providing a service locally. All of this movement is triggered with ease. The ease is facilitated by an intricate web of international exchange that fulfills the demands of the global market while providing the means for products to be sold and shipped anywhere. Nowhere within this process is the architect, the builder, the stone supplier or manufacturer, or anyone else involved in the project asked to assess the process for its expenditure of energy or any other resource, apart from those with a monetary cost attached. At no point in the process is the total cost of the decision to specify the particular Brazilian stone questioned on the merits of its resource footprint.

In addition, the architect, the builder, the stone supplier, and again, everyone else involved in the process do not have the tools or the training to make any kind of assessment in the selection in any other way than financial cost. In monetary terms, the various stakeholders in the process are very well trained. Each has a very

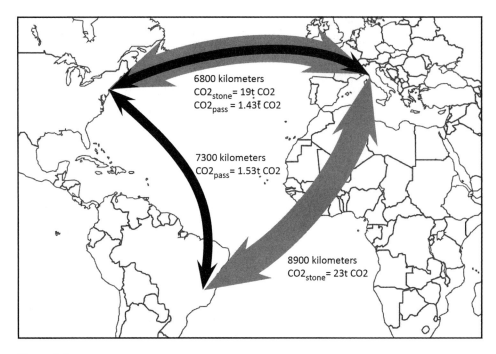

Figure 2.1
Travel expenditures as part of the process from extraction to installation for dimension stone for a building in Manhattan. CO_2 emissions are shown for the transport of three 22 metric ton, 8 m³ blocks of stone (density of 2.75 grams per cm³) from Brazil to Italy to New York City and assuming 40 grams CO_2 emissions per metric ton-kilometer transported by sea. CO_2 emissions are shown for airline passenger travel assuming 210 grams CO_2 emissions per passenger-kilometer traveled. These figures are for illustrative purposes and are approximate and conservative calculations.

precisely defined and critical stake in understanding and negotiating the price of the stone. In contrast, the stakes for resource-consumption expenditures are not even visible.

Without arriving at a precise measure of the resource consumption accompanying the decision to specify Brazilian stone, it is obvious that enormous expenditures are required. Figure 2.1 gives a general idea of the direct and indirect expenditures from travel alone, for extraction through installation.

The decision to specify a Brazilian stone primarily for aesthetic purposes engenders a total emission of CO_2 on the order of 45 metric tons; a full 68 percent of the weight of the stone itself!

This late-twentieth-century scenario—choosing stone sourced many thousands of kilometers from its use, the processing on a second continent, and final delivery to yet another continent—seeks to illustrate the ease with which contemporary global economies facilitate diverse and intense consumption. This vignette is also intended to illustrate one way of the many, many ways in which our cities are the products of global transfers—acting as the gateways for international trade. Extraction from every continent feeds our contemporary cities by way of international freight, transfer of instruments of financial value, information, and—importantly—people within the global marketplace in every sector. Today, much of this is brought to the consumer by way of the Internet and airline travel.

Constructing, operating and maintaining, and decommissioning and demolishing the built environment accounts for the majority share of the material throughput serving contemporary society. This fact was noted many years ago by pioneering researchers such as Jesse Ausubel, Robert Ayers, and Brad Allenby, whose perspective of societal and industrial metabolism made clear the enormous environmental and ecological footprint of modern construction activities. They and others have noted that upward of 70 percent of the total materials that flow through society (by weight) and 20 to 30 percent of the societal waste stream are accounted for in activities directly or indirectly related to construction (Ausubel and Sladovich 1989; Ayers and Simonis 1994; Allenby 1999; Wernick et al. 1997). While most of the volume of construction materials is nontoxic, the sheer amount extracted (along with the materials displaced in the process) makes the construction industry one of the primary drivers of environmental degradation and resource depletion. It is also linked to a vast array of non-construction sectors of the economy. For example, in the United States every $1,000 of demand for ready-mix cement concrete creates almost $1,200 of associated demand in non-construction industries, and significantly, 1 metric ton of the cement component of that ready-mix concrete requires approximately 1.5 metric tons of several other materials (Horvath 2004).

The global construction industry is a fragmented sector of the economy, both because it is geographically widely distributed and located for maximum efficiency and profitability in local markets, and because it is composed of many hundreds of thousands of very small business firms. Therefore, the dissipation of wastes from construction originates from many production and processing facilities located in many locations in every country. For example, China has several thousand cement plants distributed widely through every region of the country, each a source of airborne particulate matter and carbon dioxide emissions.

Construction is but one economic sector that serves contemporary urban places. Construction adds to the stock of the city by expanding transportation, water and power infrastructure, making buildings, managing waterways, and shaping and managing the natural landscape. Economically healthy cities have a mass balance in which the volume of wastes dissipated is always less than the volume of imported materials. The difference is found in the enormous volume of material used in adding to the long-lived stock of the city: its highways and buildings, power grid and water system.

Contemporary cities are often cited as the primary engines of consumption in our societies. Many writers, thinkers, and researchers have identified and detailed the many ways in which the contemporary city promotes widespread extraction of materials and dissipation of wastes, regionally and internationally. Many of these same researchers have sounded an alarm about the coming catastrophe of resource scarcity and regional and national conflict in some measure prompted by the inordinate demands of urban areas for critical materials, energy, and water. Agglomerating human activities and settling large populations in concentrated spaces has been cited as energy-inefficient and material-intensive compared to a natural ecology. H. T. Odum has proposed that cities consume 10 to 100 times the energy of a typical unmanaged ecosystem (Odum 1971). Urban centers are also identified as the most prolific emitters of that paramount residue of modern industrial overshoot: carbon emissions; (Intergovernmental Panel on Climate Change 2007) "Cities are the defining artifacts of civilization, but they are also dangerous parasites, with a capacity to harm regions far beyond their own boundaries. The ecological impact of cities on the global environment is out of proportion to their size." Figures 2.2 and 2.3 provide a schematic illustration of the localized impacts of urbanization.

In contrast, researchers, designers, and engineers, and others have noted that density can save us from ourselves. Clearly, having a resident population (workers)

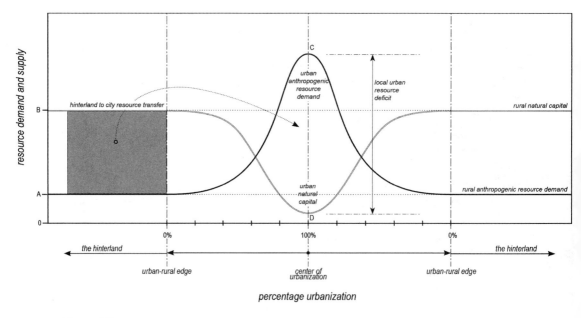

Figure 2.2
Relation between urban density and productive capacity of land.

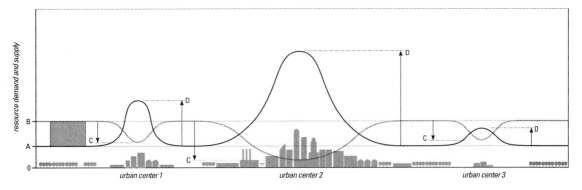

Figure 2.3
Relation between urban density and productive capacity of land within a polyurban region. A, level of resource demands in a rural context; B, level of productive capacity of rural land; C, reduction in productive capacity of land at center of urban zones; D, increase in resource demands at center of urban zones.

within the immediate vicinity of employment centers (firms) has been a timeless agglomerating force that has catalyzed and maintained the creation and proliferation of cities in every culture, on every continent throughout human history. More generally, a place that maximizes benefits while reducing costs for both people and businesses by reducing the distance between them will compel the initiation of human settlements and the growth of cities. Despite predictions at the beginning of the digital age that the combination of a dematerialized and service-based economy and the reduced need for face-to-face interactions would render agglomerating forces impotent, the opposite seems to be the result. Innovation within cities, the adjacency of "idea" centers, the heightened value of the high quality face to face interaction that cities provide seem to be a new way in which urban economic agglomeration is accelerating and producing a "gentrified *polis* of tomorrow" (Glaeser 1998, 139; Ford 2008).

Density is generally regarded as one of the hallmarks of cities—in fact, a defining attribute of urban spaces. Greater urban density is considered a positive move toward using less space, less transportation energy, less material and energy overall than cities that are spread out to cover more acreage in low-density, detached buildings.

Yet contemporary cities, even the most developed, may not naturally tend toward greater density. Numerous European examples tell us that the urban population may tend to migrate away from the solidified center. Reasons for this include the explosion of urban affluence and skyrocketing land prices, preferences for the detached housing to be found on the perimeter, location of employment on the periphery, and lack of housing in the center, among others. Many European cities now have centers that, while beautiful and unique, lack the diversity of population and urban life that created them in the first place.

Recent work shows that Manhattan was once much more densely populated than it is today. Despite the fact that the daily population essentially doubles with the commuting work force, Manhattan is home to many fewer people than it used to be. The population of Manhattan in 2012 is 1.6 million, compared with a 1950 population of 2 million and a 1910 population of 2.3 million. That's a staggering reduction of 30 percent of the island's residents in 102 years. Of course, back in 1910, 90,000 windowless rooms were used as bedrooms, and immigrants routinely found themselves in small and overcrowded tenement apartments (O'Leary 2012).

Co-location of people and employment may also be the primary anchor for urban resource efficiency. Reducing emissions due to automobile transportation, maximizing mass transit and alternative modes of urban mobility, creating large-scale, high-performance residential and commercial buildings, producing symbiotic material and energy relationships by harvesting waste heat and materials, and doing this within the confines of limited space bode well for increasing awareness of resources and the environment. While it is clear that cities have always been a major driver of anthropogenic resource extraction, consumption, and waste production, the twentieth century offers beguiling evidence that urbanization is both cause and cure for continuing compromise of the environment and depletion of critical resources. Recently, an important body of work has emerged that gives us cause to welcome the intensity and acceleration of the recent and continuing urbanization of the twenty-first century (Glaeser 2011a, 2011b, 2011c). Improved health, greater access to education and markets, and convergence of skills, knowledge, and technology, among many other elements, have resulted in extraordinary growth in standards of living and creation of wealth in cities of widespread and contrasting geographies and diverse economies and cultures. As centers for the cultivation of innovation, technology development, progressive organizations, and political engagement, cities present themselves as incubators that both require ever-more-intense convergence of critical resources while offering novel pathways toward a productive and humane future.

Measuring and Assessing Urban Consumption

How might that future address the environmental and ecological challenges facing human society? The focus of this book is the relationship between resources and urbanization. Today, one measure of the move toward some better future is the formulation of practical strategies for reducing the emission of anthropogenic carbon. A widespread goal is the notion of carbon-neutral consumption. Carbon-neutral airline travel, rental cars, homes, and office buildings, even carbon-neutral cities, islands, and regions are receiving a great deal of attention. Alongside Dubai in the United Arab Emirates, Masdar City is rising with a stated goal of operating as a carbon-neutral city—the first of its kind if it proves successful (more on Masdar later in the book).

How do ideas of carbon neutrality, net-zero energy buildings, urban water consumption efficiency, material recycling, and downcycling match the reality of global resource consumption? How can we begin to assure ourselves that these emerging and changing aspirations have any hope of fulfillment? In the twenty-first century, how does the kind of consumption illustrated by the architect choosing stone, an engineer specifying a polymer carbon fiber composite for an airline wing, a bride ordering exotic flowers, or a Fortune 500 company exercising a mining concession—how does all of this activity add up in terms of global material and energy resource flows?

The world has seen many enormous increases in resource consumption. The various phases of economic and social expansion accompanying the rise and fall of many great civilizations are not often viewed through the prism of their enormous consumption and significant effects on the environment and all manner of resources (Ponting 1991; figure 2.4). From the earliest establishment of human settlements—truly sedentary enclaves of people—environmental degradation and resource depletion

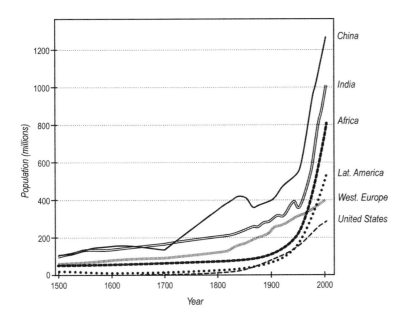

Figure 2.4
Population growth over the last 500 years in China, India, Africa, Latin America, Western Europe, and the United States (adapted from Maddison 2007).

have been an ongoing reality. Even in the earliest urban settlements, in Mesopotamia and later in Greece and the Roman Empire, rampant deforestation resulted from the construction of buildings and ships used for trade and war. The use of wood as a fuel, especially for the firing of masonry units made of earth, consumed vast volumes of wood and led to widespread deforestation in ancient Jordan, Iraq, and other parts of the Middle East and Mediterranean. Another example is the rapid rise in the European population from after the Black Plague well into the nineteenth century, especially during the advent of the Industrial Revolution. The need for cleared agricultural land marked the end of many ancient forests during the early medieval period from about 1,000 A.D. until the dramatic decrease in population following the Black Plague of the mid-fourteenth century. And the devastation of forests has continued well into recent centuries and today. Ponting gives the example of Ethiopia, when Addis Ababa became the capital in 1883. Only twenty years later, forests for many miles surrounding the city were completely cleared—a result of making charcoal for urban household fuel, an issue even today as urban populations continue to grow (see figure 2.4).

It has been conclusively shown that developed economies tend toward less materials use per unit of gross domestic product (GDP), accompanied by greater per capita resource intensity. That is, personal and household affluence has a tendency to accelerate per capita resource consumption, while economic expansion can proceed at a more rapid pace than the increase in resource use. The specter of a similar decoupling between materials and energy consumption per capita and overall economic growth presents the urban world with a dilemma.

Urban contexts tend to increase affluence, while attracting rural populations. Agglomeration economies work very well in attracting a skilled work force that proceeds to innovate and create wealth. This creation of intellectual capital drives greater worker productivity but also creates ever more intensive needs for materials and energy—especially nonrenewables that can be used to build the sophisticated information and urban infrastructure required of these knowledge centers.

Recent studies have assessed the twentieth century in just this way. It is now clear that the twentieth century not only hosted an explosion of the global economy and population, but also saw significant increases in the resource intensity per capita. Data for the twentieth century clearly show consumption at an alarming rate in quantities never before mobilized and of every kind of material and fuel. The twentieth century was truly exceptional in the history of humankind in the rate at which the economy dipped into global resources to fuel activities of every kind (McNeill 2000). The century was also unprecedented in the diversity and amounts of waste materials, from carbon to PCBs to pharmaceuticals, that were spewed into the global environment.

To understand the enormity of the increase in consumption, it is useful to focus on extraction—the phase in the materials cycle that includes the various processes employed to remove and refine minerals, biomass, water, and other materials for use in the global economy (Krausmann et al. 2009). Mining of minerals and fossil fuel precursors (or carriers) like crude oil, logging, agriculture, and many other activities are employed globally in extraction, the first phase of consumption.

Global extraction of materials served an increase in the population to 6.4 billion people—a 400 percent increase since 1900. At the same time, the global economy, as measured by a summation of national gross domestic product expanded more than twenty times.

The increase in materials extraction during the twentieth century occurred in roughly three phases. The first lasted from 1900 until after World War II and resulted in only a modest increase in overall direct material consumption (DMC). The second phase saw a doubling of the annual DMC, partially prompted by the reconstruction effort in Europe and Asia and the expansion of economies of the world, especially that of the United States. This growth was substantially muted for a few years during the energy crisis of the 1970s. The third period leading to the beginning of the twenty-first century has seen an acceleration of the rate of extraction and consumption of materials worldwide (Krausmann et al. 2009).

A key finding shows that DMC growth outpaced population during the twentieth century. Therefore, material intensity per capita increased—in fact, it doubled during these one hundred years. Each person today consumes about twice that of a person living in 1900. Therefore, the actual physical economy grew more rapidly than the global population. In contrast, the economy outpaced even this rapid rise in materials extraction. The material intensity of the global economy of year 2000 decreased by more than half 1900 levels. As a result, each unit of GDP today requires less material input than in 1900, and it is estimated that between 47 and 59 billion tons of material of all kinds were extracted in the year 2000 (Krausmann et al. 2009; Sustainable Europe Research Institute 2008). This is eight times the amount extracted at the beginning of the twentieth century. Also, the greatest increase in materials extraction was in construction materials used to expand infrastructure and build the urban fabric needed to house the vast increase in city populations and the increasing concentration of urban industrial and employment centers. The extraction of construction materials increased thirty-four times over 1900 levels.

In addition, the projections for the increase in urban populations bodes more of the same for many decades to come. Today, materials and energy use in developing regions is below the global average; however, per capita consumption has begun to accelerate, and the overall growth of DMC is beginning to outpace population growth even in these developing regions of the world.

The Global Sociometabolic Transition

It is clear that the world is fully engaged in, and in some areas has completed, the transition from an agricultural sociometabolic regime to a fossil fuel regime. Urbanization is a very good proxy for this transition. However, the distribution of this transition globally is not a smooth one; it is important to acknowledge the differences in rates and levels of urbanization between different regions and countries in the world. The developed world is already substantially urbanized, at around 70 percent, much more so than the developing world, which remains at roughly 40 percent in Asia and Africa. Trends today and for the foreseeable future indicate that developed regions have reached maximum urbanization or may de-urbanize slightly in the coming decades. During the 1990s, 40 percent of the cities in the developed world experienced population contraction, while approximately half grew at a very small rate of 1 percent per year. Cities in the developing world grew at much faster rates; 17 percent grew at an extremely fast rate of 4 percent or more, and 36 percent at rates between 2 and 4 percent (United Nations-Habitat 2008).

During the twentieth century, urban populations increased like never before. Today, urbanization adds approximately 193,000 people to cities every day. While the rate of growth of the urban population is now slowing, it will continue to drive massive population increases in many regions of the world. The rate of urban population growth between 1950 and into the first decade of the twenty-first century averaged 2.6 percent per year, though projections for global urban population growth decrease to 1.8 percent, and then eventually 1.3 percent during the second quarter of the new century.

This has driven the size of the global urban population to historical highs in much of the world. Latin America and the Caribbean count 78.3 percent of their populations as urban, while North America has reached over 81 percent, and Europe over 72 percent. This is the result of a century of urbanization that will continue for several decades to come. The global urban population in 1950 amounted to 740 million people (30 percent of the total 2.54 billion people) and is projected to increase to 6.4 billion in 2050 (70 percent of the projected total of 9.19 billion).

Urban population growth rates in Asia and Africa will persist above 1 percent per year through 2050. During the period 1950–1975, African urbanization exceeded that of Asia, at 2.28 versus 1.42 percent. This was reversed between 1975 and 2007; Asia urbanized at a rate of 1.66 percent compared to 1.28 percent for Africa. The percentage of the population in Africa and Asia will remain closely comparable; today both are urbanized to approximately 40 percent of their populations and are predicted to reach 62 percent in Africa and 66 percent in Asia (figure 2.5).

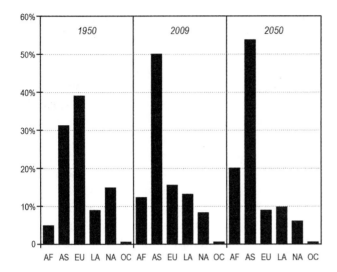

Figure 2.5
Urban population levels showing distribution by region; AF, Africa; AS, Asia; EU, Europe; LA, Latin America and the Caribbean; NA, North America; OC, Oceania (adapted from United Nations 2009).

The absolute numbers tell a more sobering story. Even though the urban percentages in Asia and Africa are both approximately 40 percent, Asian cities now account for half of the world's urban population, while African cities only amount to 11 percent. The total urban population projected to be living in Asian cities in 2050 is 3.5 billion people. In Africa the number will amount to a little less than one third of this, at 1.2 billion. The total urban populations of Latin America and the Caribbean, North America, and Europe combined in 2050 will amount to approximately 1.6 billion. Therefore, the Asian urban population will exceed the urban population of the rest of the world by 600 million people—currently the entire (urban and rural) population of Latin America, the Caribbean, and Oceania combined (figure 2.5).

Today, and for the coming decades, the majority of the world's urban population will continue to reside in a small number of countries; 75 percent are found in only twenty-five countries. In fact three countries, the United States, India, and China, currently account for 35 percent of the global urban population. The future will continue to see further concentration in just a small number of countries, though some countries with relatively low levels of urbanization will rapidly increase their city populations. These include Nigeria, China, Bangladesh, Pakistan, Indonesia, and others.

In addition, several databases and studies show clearly that the composition of the material economy has steadily and strongly shifted from domination by biomass to greatly increasing concentrations of minerals. In fact, it has been suggested that economic development the world over, increasing urbanization, and the acceleration in the construction of infrastructure has led to a global economy much less dependent on materials that cycle through the economy quickly (such as agricultural food products, wood for fuel, and other biomass materials) to metallic and non-metallic minerals that tend to accumulate in the built stock of society (Hashimoto, Tanikawa, and Moriguchi 2007). That is, biomass is more directly tied to human nutrition, while nonrenewable minerals are employed in economic expansion (Krausmann et al. 2009). This addition to long-lived durable stocks locks in an increased capacity for energy consumption for decades to come. Urbanization is a good measure of this transition away from biomass consumption, since urban areas today are very much dependent on an economy fueled by fossil energy carriers. Of course, the urban world can only be counted as one driver of this increase in resource extraction and consumption. However, it is clear that the urban world is a major, if not primary, driver for the past and future increases.

Three Phases of Urbanization

Cities are material-intensive. Large volumes of material are mobilized, processed, and placed to serve the large-scale infrastructure needs of the urban population. Buildings require many tons of rock, gravel and other aggregate, cement, steel, brick and tile, bitumen, paper, gypsum, and many other materials. This material input is required to mobilize critical resources (for example, water, food, and energy) that are required for the operation of the city and the sustaining of the urban population over long periods of time.

In the schematic way shown in figure 2.6, cities undergo three phases of urbanization, labeled as phases 1–3 below. The fourth phase is that which is anticipated and suggests a "sustainable" urban future.

Phase 1 initiates the materialization of the city as the basic framework is laid out. Streets and building plots are surveyed and established, networks of infrastructure for distributing power and water and collecting and carrying away wastes are engineered, and construction begins. This phase entails the input of enormous quantities of materials. Chief among these are gravel and sands associated with shaping the landscape and providing the basis for urban construction.

Phase 2 begins as the intensity of materialization begins to decrease. This may be the result of a slowing down of either the spatial expansion of the city or the solidification of the urban core, or both. In any case, a slowdown of additions to stock does not mean that materialization will ever go to zero. Existing stock undergoes constant maintenance and repair requiring material inputs that sometimes begin to approximate the intensity of phase 1. For example, today the majority of

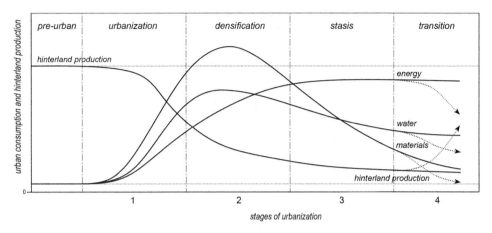

Figure 2.6
Four stages of urbanization and associated material and energy consumption. Phase 1 encompasses the initiation of urbanization; phase 2, the densification of the urban fabric; phase 3, the slowing of the rate of urban expansion and population increase; and phase 4, the much-anticipated transition to a sustainable future through fundamental structural shifts in the acquisition and distribution of critical urban resources, such as water, energy, and food (adapted and expanded from Fernández 2007).

the material throughput due to construction on the island of Manhattan is caused by the material requirements of reconfiguration and renovation of interior residential and commercial spaces within existing buildings. Generally though, the material input will likely level off and decrease as the major additions to the infrastructure and building stock are completed.

However, as this occurs the consumption of energy continues to increase. As the city matures, the population increases and becomes wealthier, consumption focuses on energy-intensive appliances like air conditioning, electronic devices, and other products, and household energy use steadily rises. During the privatization of the housing market in China in the 1990s, household electricity use exploded in Shanghai and other increasingly affluent cities. This is a hallmark of phase 2: materialization may slow down, but energy use will continue to increase.

Phase 3 may occur early in a city's history, within decades of its founding, or it may not commence until after several hundred years of existence. This phase marks the incremental densification of much of its urban fabric, ultimately ending at a point of stasis in which the built environment and the density of infrastructure have reached a terminal point. This is often the result of a land use end condition in which size and density are predetermined and regulated. A zoning code will act to establish ultimate densification as setbacks from property lines are finalized, built volume to open space proportions is determined, and maximum populations are reached. During this final stage, materialization decreases to levels that just satisfy a steady flow of material resources for maintenance and repair. As the city becomes more service-oriented and the population becomes ever more affluent the city may incrementally increase the density of the built environment, while the density of the population may actually decrease as households expand and amenities demand more interior space. During this phase, the rate of increase in energy consumption may slow, but it is unlikely that absolute energy consumption will decrease. Energy consumption will continue to increase unless the urban economy falters, employment falls, and a net migration out of the city leads to a substantial shrinkage in the urban population.

Phase 4 is the transition in sociometabolism in which the solar economy is fully realized. In an urban context this suggests a structural shift in the urban infrastructure such that resources, materials, and energy are acquired, organized, distributed, and consumed in ways that maximize the closing of material loops and the harvesting of substantial, if not comprehensive, renewable energy sources such that urban demands are met. Thus, phase 4 remains an aspiration.

Aging Urban Populations

An important aspect of the changes occurring in our urban world is the shifting demographics underpinning societal metabolism. In most countries today, demographic trends are clear. Populations are growing older. Decreasing birth rates, increasing life expectancies, government policies limiting births (notably in China), better access to education, and growing global affluence are leading to a global population that is older than ever before. In 1950 8 percent of the global population was 60 or older; today it has increased to 11 percent, and is projected to increase to 22 percent by 2050. Given overall population projections, 2 billion people will be 60 years old or older by 2050 (United Nations 2009).

The consequences for urban contexts are varied. On the one hand, older people are generally less economically productive than their younger employed counterparts, and they also incur greater health care costs. However, older people tend toward lower overall consumption and are less prone to engage in risky and criminal behavior. Also, urban places offer ease of access to health care, markets, and cultural and other activities that significantly enliven the lives of older people. Communities that are planned to serve the special needs of the aging community may play a special role in the evolution of urban form and infrastructure. More diverse and reliable public transportation options are an important element that would benefit from urban residents who cannot or choose not to drive.

Future urban development would do well to consider the various opportunities and challenges for engaging the lives of the older segments of the population, as this will be a trend for many decades to come.

Urban Activities and an Urban Metabolism Framework

We can begin to examine the resource consumption of cities by considering the varied activities that are contained by and supported by cities. Urban activities are a particularly powerful method for distinguishing between the myriad kinds of consumption in the city. Why is this so?

First, correlating consumption with broad sets of activities ensures that a systems approach can be applied to both analysis and design. This means that the correlation may engender a concise and generic formulation of urban resource consumption. It is true that cross-sectoral consumption is an inevitable outcome and one whose comprehensive understanding continues to elude even the most advanced research. However, activity-based notions of urban consumption are not spatially generic. Activities occur in locations, and delineating areas in which activities take place and specific locations for those activities is essential. Second, basing assessments of resource consumption on activities rather than economic sectors places the primacy of the analysis on the flow of physical resources, not financial value. Third, carefully designating a classification system of essential urban activities links urban resource studies with research in other disciplines, especially urban economics. Fourth, and possibly most important, activities constitute the life and soul of urban places. Urban concentration (economies of agglomeration) occurs to facilitate activities that would not otherwise occur. Transactions of every kind occurring at an enormous range of physical and temporal scales both define urban places and are defined by the urban context.

However, this book addresses urban activities at a meta-sectoral level. That is, as used here, *urban activities* refer to the aggregate consumption related to a sector, a district, and a fundamental provision of the city or other grouping of socioeconomic actions that involve resources. Therefore, urban activities are collections of actions devoted to arriving at a particular result. Several classifications of these kinds of urban activities are commonly used.

The framework presented here delineates energy and material flows devoted to three broadly inclusive sets of urban activities (see figure 2.3):

1. the provision of habitable space (the built environment, ua_2);
2. the provision of goods and services of all types (products, ua_1); and
3. the provision of the movement of goods and people (transportation, ua_3).

These urban activities are formulated as *provisions* of urban living and working. That is, the city is conceived of as a collection of necessary and sufficient provisions of habitable space, goods and services (especially air, water, food, critical materials and waste removal), and transportation.

This formalization is intended to provide a robust intellectual and operational link to the main theoretical assertions of economics and urban ecology. Specifically,

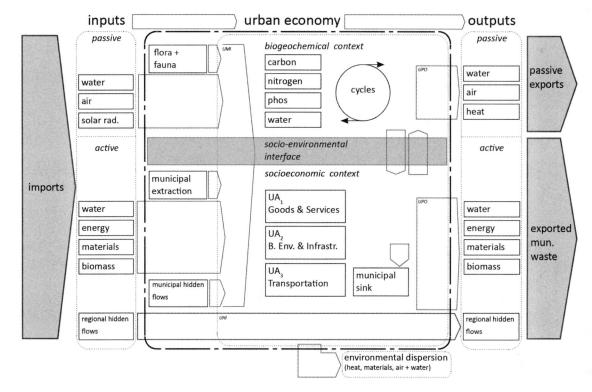

Figure 2.7
Urban metabolism framework.

spatial equilibrium is organized according to the production of firms and workers (goods and services) and the costs of housing (built environment) and transportation. These explicit links lend important guidance in the ongoing project to connect economic models of urban growth and development with models of resource consumption that take into account natural cycles and biogeochemical processes, as shown in figure 2.7.

The modeling of urban systems and processes has been carried out on a range of levels by a diverse set of researchers from many disciplines. Simplified models are able to capture complex urban scenarios in a useful way. The development of a simplified model of urban metabolism can illustrate:

1. the bulk of energy and material flows prompted by urban socioeconomic activities;

2. the overall relationships that characterize urban biogeochemical processes;

3. primary interactions between socioeconomic and biogeochemical activities, processes and cycles within the architecture of an integrated stock and flow model coupled to a system dynamics module; and

4. reasonable results that provide guidance in delineating critical positive or negative behaviors of the urban zone.

In addition, there are seven distinct reasons why simplified models may play a useful role in the context of the modeling environment and urban studies needs of today. The development of a simplified model of urban metabolism can provide:

1. overall direction to planners, engineers, designers, and policymakers in the general assessment of the resource consumption behavior of urban zones;

2. overall understanding of the interaction between urban socioeconomic and biogeochemical processes;

3. accessible and agile tools for the assessment of a diverse range of urban resource issues;

4. accessible and agile tools to aid in the early development of green city initiatives;

5. insights into aspects of the fundamental resource consumption behavior of urban zones;

6. direction in the development of generalizing principles upon which a wide variety of cities can be modeled and assessed; and

7. direction in the development of a general typological scheme of cities based on distinct resource consumption profiles.

More work is required to develop consensus around the components of these kinds of simplified models, and this will constitute a major development to be expected in the urban metabolism field in the near future.

Several models that attempt to describe resource flows within the economic and ecological context of urban regions have been developed (notably UrbanSim/OPUS). The use of these models for detailed analysis of specific cities and their regions is well established. However, robust simple models that generalize dynamics of urban systems based on an accounting of resource flows are not well developed. The use of these models as avenues toward better understanding of urban typologies and characteristic urban resource consumption profiles is needed.

It is important to note that the notion of urban activities used in this chapter is at the mesoscopic and macroscopic levels, not the microscopic. This is an important clarification in light of the fact that the phrase *urban activities* is often used today when referring to the behavior of individual urban dwellers. Today there are an increasing number of efforts that intend to track the activities of individual urban dwellers within the urban spatial context. Recently, tools to acquire data at very finely resolved detail have made it possible to launch these kinds of efforts with the prospect of learning, for the first time, about the actual consumption and travel patterns of large segments of the urban population. Typically, this has been done using data gathered from information and communication technology networks and devices. Targeting mobile phone activity is now a well-known strategy for gathering this kind of urban agent activity.

3

Intellectual Foundations and Key Insights

The effort to provide a comprehensive understanding of the contemporary city has understandably become highly multidisciplinary. This is a positive development for an area of study that requires ecological, environmental, and economic studies informed by a vast array of subdisciplines and distinct methodologies. Given this intellectual diversity, endeavoring to understand the current state of progress toward sustainable cities is challenging. Therefore, it is useful to survey the disciplines currently involved in the study of cities directly and indirectly and examine their respective interests as well as the relationships between them. This chapter undertakes the task of identifying those fields most relevant to the urban metabolism effort and provides guidance in acquiring an understanding of the state of the notion of urban metabolism as defined by the diverse work of many researchers. Each field discussed in this chapter is briefly outlined in terms of its relation to the area of urban metabolism and key contributions already achieved or possible within the short term.

An Intellectual History with a Future

To experience a city is a daily routine for most of humanity. Most people today spend some part of their day living or working in cities and sharing in the urban experience. In fact, it is so common to know of the joys of a walk through a favorite neighborhood, to shop along an urban retail boulevard or to contend with the frustration of congested traffic encountered commuting into and out of an urban business district, that it seems we rarely question the fundamental workings and efficacy of this dynamic complex system. Cities, and the way they work, seem inevitable consequences of human civilization—inevitable as the view of Chicago in figure 3.1 implies. That they may become the essential element of a new and sustainable civilization intrigues many.

Many sectors of contemporary society, including the academy, business, and government, are actively contemplating the nature of urban systems and the costs

Figure 3.1
The city of Chicago (photo J. Fernández).

and benefits entailed in delivering vital resources to serve fundamental urban activities. This has led to a great deal of new work regarding every aspect of the form and function of urban systems and the socioeconomic and biogeochemical consequences of our urban era. From a multitude of peer-reviewed journals, newly published books of all kinds, green city plans, and reports from think tanks and government agencies at all levels, the proliferation of work regarding the city can be considered evidence of a resurgence in urban studies as a central concern of researchers, designers, engineers, and policy makers. Driving this flurry of activity is an intense interest to better understand the nature of the contemporary urban fabric and the associations with critical resource constraints (water, energy, food, and clean air), global climate change, and societal priorities such as national and regional security and resilience.

However, who is charged with addressing these issues? Who is to consider the city in all its complexity for the purpose of proposing alternative modes of production and consumption acceptable to and economical in an urban world? What set of experts and what parts of society are best suited to addressing the urban future? Today, the sheer number and diversity of questions underlying the drive toward sustainable practices requires a veritable army of researchers and workers striving to provide new perspectives, productive tools, and effective policies to implement change in the urban world.

Answering these questions requires that we identify and describe the work that is coming from and informing a variety of related and unrelated intellectual, business, and policy-oriented pursuits. Therefore, the primary purpose of this chapter is to survey the current state of the emerging intellectual foundation of urban metabolism primarily in terms of the various fields that are engaged in the multifaceted work required to arrive at a sustainable urban future. It is through these fields that the complex effort to understand cities and their relation to resources and the environment is being undertaken. From industrial and urban ecology to urban and ecological economics, from research centers to the offices of national, regional, and municipal governments around the world, the aspiration to embark on a more sustainable urban future is proceeding with the allure of positive transformation.

In addition, this chapter attempts to outline the overlapping interests and priorities of distinct fields. Though many of the methods and tools under development are quite different, many of the goals and underlying motivations are surprisingly similar. However, each field is under pressure to include useful information from other fields as it delineates its own intellectual territory in ways that optimize the possibility of success in reaching specific goals. As a prominent urban economist writes, "While I believe that no one can make sense of cities without the tools of economics, I also believe that no economist can make sense of cities without borrowing heavily from other disciplines" (Glaeser 2008). This observation applies to all fields hoping to contribute to a better understanding of the city.

Another goal of this chapter is a listing of the most important outstanding research questions relevant to urban metabolism in each highlighted field. Outlining these questions and their relation to questions emerging from other associated fields will assist to elucidate the interdisciplinary landscape of urban metabolism today and in the future. Several fields share key questions, while their approaches and research goals may be quite different.

Finally, this chapter places the work of these fields in an important causal framework that relates societal metabolism to environmental consequences. Adopted by the European Environment Agency and introduced in chapter 2, the DPSIR framework provides a way to conceptualize the socioeconomic drivers, pressures on, and state of the environment, as well as environmental impacts on and responses by society. Relating each of the highlighted fields of urban study to this guiding framework fosters implications for coordinating work in one field with that of another, even a completely unrelated and contrasting field.

Topics of Common Interest

It will become quite clear in this chapter that the effort toward a sustainable urban future is far from coordinated. In fact, it may only be held together by a general notion of the need for contributing to managing the operation and growth of existing cities, while assisting with the emergence of new cities in humane and sustainable ways. The hope for coordinated action is often left unfulfilled. Despite this array of uncoordinated investigations, it is possible to articulate a short list of four key topics that seem to be forming an overarching intellectual framework for the research that is coming together to define urban metabolism. The broad interest in and intellectual appeal of these four topics provides a strong rationale explaining the diversity of fields engaged in the emerging formulation of urban metabolism.

The first of these topics is a set of ideas and associated methodologies that define the notion of societal metabolism. Originally arising from biology and then ecology, the application of the notion of metabolism to the respiration of a system composed of many organisms acting in a complex spatial and temporal context has resulted in a rich conceptual and methodological framework. The natural ecologist E.P. Odum is one of the first to have extended the idea of metabolism beyond the biology of individual organisms and applied it to a study of the *system* of the habitat and natural ecology within which diverse species live out their lives (Odum 1953). Odum emphasized the relation between elements of a system and used this understanding to propose that consumption run amok could endanger human society:

"The basic problem facing organized society today boils down to determining in some objective manner when we are getting *too much of a good thing*. This is a completely new challenge to mankind because, up until now, he has had to be concerned largely with too little rather than too much. Thus, concrete is a *good thing*, but not if half the world is covered with it."

He continues in the same article,

"Society needs, and must find as quickly as possible, a way to deal with the landscape as a whole, so that manipulative skills (that is, technology) will not run too far ahead of our understanding of the impact of change."

He goes on to conclude,

"[we] have not yet risen to the challenge of the urban-rural landscape, where lie today's most serious problems." (Odum, 267; 1969).

The fact that a natural ecologist includes the urban–rural landscape as he advocates formulating "some objective manner" to analyze our consumption illustrates the central place that the urban context has held in matters of sustainability. The idea that cities, as distinct entities of physical concentration and economic, technical, and cultural agglomeration, can be analyzed and assessed in terms of system behavior and dynamic resource flows is an easy outgrowth of notions of societal metabolism, which in turn owe at least part of their origins to work in natural ecology (Fischer-Kowalski 1998; Fischer-Kowalski and Hüttler 1999). Simply put, urban metabolism is a subset of societal metabolism.

The second topic is an understanding of the almost complete domination of the earth's vast array of biomes and exploitation of every useful mineral and biological resource. For many years now there has been a clear determination that the transformational effects of anthropogenic activities can be found in every biome on earth (for example, Vitousek et al. 1997). Recently, a proliferation of studies that characterize global resource flows has led to a better understanding of the network of material and energy exchanges that serve contemporary society and the ramifications for the environment and human society (Schandl and Eisenmenger 2006; Behrens et al. 2007; Weisz and Schandl 2008). For the first time, large-scale studies are able to trace the extraction and consumption of the world's resources during much of the period of human history since, and sometimes before, the Industrial Revolution (Erb et al. 2008; Krausmann et al. 2008). This has led to the proposal to acknowledge entering into a distinct geological period called the *anthropocene*, a period defined by the primacy of human activities in altering biogeochemical cycles to an extent that planetary systems are beginning to change state from that of the Holocene epoch to something new.

The third topic framing this chapter is a growing awareness that the evolution of economic value in the urban context and the associated impacts on the regional and global environment serving that context are inextricably linked. The flow of materials and energy, emissions of particulates and greenhouse gases, and concentration of nutrients and critical stocks in cities serves as a clear signal that urban economies and global sustainability are coupled together in many complex ways. There is nothing new in this idea, except for the yet-to-be-accomplished merging of classical and ecological economics in a mutually consistent and productive manner. How will this be done? While it may be obvious to state that our socioeconomic structures are physically coupled with and (for the most part) determine the health of the environment, it is not obvious what the prospects are for a productive integration between financial markets, economic valuation, resource flows, and environmental impacts. The adoption of environmental and ecological concerns and the proclamation of urban sustainability goals by municipal governments combine to alter in some fundamental ways the notion of urban economic development. Creating value in this new urban context is an ongoing experiment in many enlightened cities.

Finally, the fourth topic is the adoption of alternative disciplinary structures and processes that serve the critical need for integration studies. While this call for multidisciplinary cooperation and integration is heard in many sectors of the academic world and some parts of the business and governing communities, it is particularly relevant to studies that are attempting to map out alternative paths of efficient resource consumption by large swaths of urban society. Material flow analysis and system dynamics are two methods that serve this drive toward integrated studies, but one can expect that alternative methodological approaches will continue to emerge and evolve motivated by the ultimate aspiration of providing holistic understanding of vast, complex urban systems. The questions that arise include both the mode and extent of integration between fields and the intellectual territories carved out by the development and adoption of novel methodological approaches. This fourth topic acknowledges the fact that there is no single field that completely captures the scope of methods necessary for a productive, intellectually rigorous, and integrated understanding of contemporary cities.

Currently, these four topics are related to the assessment of the resource consumption of urban areas in various direct, indirect, and associative ways. This is, as outlined in other parts of this book, a result of the novelty of the analysis of urban systems as a metabolizing whole and the application of this analysis to making sustainable cities. It is so novel, in fact, that the defining character of the work in urban metabolism is the decidedly forward-looking nature of it all. The present urban conditions in almost every city on earth do not differ radically from approaches to urban resources during most of the twentieth century and the beginning of the twenty-first. Today, it is not possible to cite many examples of real and sustained progress toward urban sustainability despite all of the rhetoric about the needs and prospects for radical improvements in the efficiency with which urban resources are consumed. This is understandable. The "lock-in" of urban respiration can be cited as a main cause of stasis. Transportation cannot be transformed overnight because highways will not be easily altered or removed. We will not immediately achieve the goal of a zero net energy and carbon-neutral built environment, because low-performing buildings can only be retrofit for higher performance slowly and in concert with the economic realities determined by local real estate markets and construction pricing. Regional and national energy production and distribution will be transformed at a pace commensurate with the enormous cost and engineering challenges of altering enormous portions of the energy infrastructure. It seems inevitable that widespread success remains in some future time to come.

Fields Contributing to Urban Metabolism

As noted above, work toward a sustainable society and resource-efficient cities is being done by an ever-increasing and often unrelated set of diverse disciplines and fields. While this is certainly necessary in order to fully address the broad array of issues that challenge our immediate urban reality, it leads to some difficulty in tracking and assessing the state of the effort. Are urban economics and architectural design converging toward the same kinds of urban sustainability goals? Do they even share common urban sustainability goals? Do the fascinating results coming from studies in urban morphology match emerging planning strategies for sustainable cities? Can results derived from the study of dynamic complex systems be applied within politically charged and socially challenging planning contexts? How much of the work in urban sustainability converges in complementary ways, and how much diverges to vastly irreconcilable outcomes? Is it inevitable that many results will remain stubbornly incommensurable?

The listing below is an attempt to map out common ground and the sometimes-opposing objectives of the work of the many fields that have identified urban resources as an important issue.

In order to do so, the following survey is organized into three groups. The first group, tier 1, describes those fields that have contributed directly to the development of strategies for sustainable cities. This group can be shown to have directly addressed the challenge of sustainable cities and brought some clarity to the approach of formulating the concepts, methods, and tools of urban metabolism. Fields included in this first group are:

- industrial ecology and environmental resource management
- urban and regional planning
- environmental research and engineering.

Tier 2 includes fields that have made indirect, though important, contributions and are considered here to be indispensable to a continuing development of the concepts, methods, and tools of urban metabolism. Of particular note in this group is the work in urban economics, which is indispensable to an understanding of the future of cities. Tier 2 consists of the following fields:

- architectural design and civil engineering
- urban ecology
- ecological economics
- urban economics and economic geography.

Finally, tier 3 comprises fields that are contributing to the topic of urban metabolism as well as the broader topic of sustainable cities and regions. While this third tier is certainly an important collection of fields contributing to urban research generally, the value of the work itself for urban metabolism is not yet entirely clear. Tier 3 includes:

- urban morphology, scaling laws and rank size
- systems dynamics
- sociology

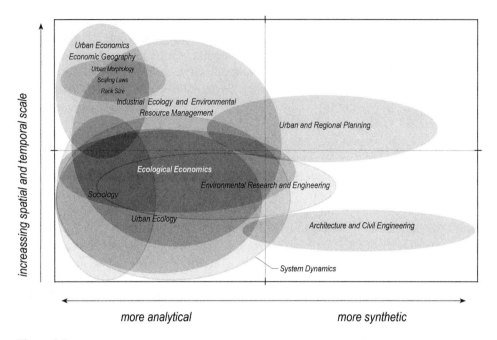

Figure 3.2
A relational diagram of the temporal and spatial coverage of various disciplines versus their emphasis
on analytical and synthetic activities. *Synthetic* is used here as an activity in which analysis is applied to
a design or technology proposal.

For the purposes of this book, distinct areas listed below referred to as *fields, disciplines*, and *domains* encompass a variety of research and professional communities, including engineering and design professions (civil engineers, planners, urban designers, and architects); research areas such as urban ecology, industrial ecology, and others; and broadly defined intellectual pursuits in sociological, economic, and historical studies. That is, this is a mixed grouping of distinct communities, identifiable research agendas, and broadly defined intellectual areas spanning a range of temporal and spatial scales as shown in figure 3.2. The intent here is to be inclusive, rather than restrictive, while being very clear about the past, current, and possible future contribution of each area. While the designation of a field may be broad and general, its relationship to the common project to foster sustainable cities will be outlined in some detail below.

Admittedly, this is a subjective hierarchy substantially colored by the relationship of each field to the emergence of urban metabolism. Therefore, the delineation between these three tiers is debatable, and the specific assignment of a field to one tier or another is subject to question depending on one's perspective. The real value of this classification is its attempt to provide a broad rendering of the intensely multidisciplinary character of the work in sustainable cities, and particularly urban metabolism. It should be noted that this survey acknowledges that the "field" of urban metabolism is a rapidly developing, though only recently emerging pursuit, based as it is on a tightly controlled metaphor. Held together by the notion of societal metabolism, this hierarchical listing should provide the reader with a better overview and productive starting point toward understanding the extent of the intellectual landscape of urban metabolism.

Tier One

The first tier includes three areas directly involved in the formulation of urban metabolism, albeit in different ways and under divergent motivational drivers. The first, industrial ecology and resource management, is highlighted in chapter 1 and is listed for its primary role in originating, developing, and promoting the conceptual framework, methods, and tools of urban metabolism. Industrial ecology itself is an inherently multidisciplinary field, composed of members from many fields listed here. Urban and regional planning is listed second because of its engagement with a broad range of topics regarding cities and their relation to regional landscapes. Finally, environmental research and engineering includes disciplines as diverse as ecology, the civil engineering specialties (in particular, water and transportation engineering), and all energy-related areas.

Industrial Ecology and Environmental Resource Management
The activities of researchers in the fields of industrial ecology and resource management provide strong evidence that this community is the logical host for urban metabolism studies. The basis for this conclusion includes the strength of the conceptual framework of industrial metabolism and industrial symbiosis, coupled with a growing list and depth of methodological approaches and tools offered by industrial ecology. Chapter 4 explores in detail how the industrial ecology toolbox is aimed at providing a framework of methods and metrics to address the variety of dimensions and length scales required to understand urban metabolism. It is the multidisciplinary set of studies produced by industrial ecologists that holds promise for capturing the greatest set of key issues in urban sustainability.

The sheer coverage of dozens of critical aspects of societal and urban metabolism—from material and energy flows to life-cycle resource requirements and environmental consequences of products, industries, and entire economic sectors—only partly qualifies industrial ecology as the most suitable host for urban sustainability studies. In addition to these methodological assets, the field of industrial ecology extends attention beyond the mere technical articulation of the inner workings of industrial production and societal consumption to a questioning of the sociotechnical framework within which contemporary production and consumption occur. Industrial ecology approaches the "understanding [of]…environmental problems in terms of how they are embedded in various contextual forces that have both shaped and been shaped by the larger industrial system being examined" (Smith 1998; Allenby 1998).

Of course, the value of methodological assets should not be underestimated. Several methods for assessing material intensities of economic activities are reviewed in this book. These methods have provided a clear path toward precisely identifying and quantifying the actual burden on the environment, draws from natural capital, consequences for natural ecological systems, and many other aspects of societal metabolism. Consider the development of the IPAT (impact = population × affluence × technology) equation. Its use as the vehicle for thought experiments is well known, and debate about its continued use is strong (Chertow 2001). It provides a pathway for productive discourse regarding the essential contributors to our environmental footprint.

This approach requires a multilevel and multidisciplinary assessment before reaching satisfying and robust conclusions that can be acted upon. In this sense, industrial ecology has been advanced as the "science of sustainability," distinguished as the nexus for directed analysis and synthesis, policy making, and design proposals. As a science of sustainability, the goals and priorities of the industrial ecology effort are determined not by the scientific community alone, but in partnership with those charged with implementation and action in a world in need of direction toward greater resource efficiency (Clark and Dickson 2003; Ehrenfeld 2004). In addition, industrial ecology serves as the primary field within which diverse methods can be brought together both to analyze the resource flows devoted to urban activities and to provide guidance for developing policy and implementing various strategies toward urban sustainability.

Environmental resource management activities are also fundamental to the study of urban resources. After all, the first paper that invoked the *metabolism* of urban systems was written by Abe Wolman, a professor at Johns Hopkins University and founder of that university's department of sanitary engineering (Wolman 1965). Wolman essentially approached the notion of urban metabolism from the perspective of the management of environmental resources, in his case primarily with regard to water. While neither industrial ecology nor environmental resource management is formalized as a *professional* discipline, each provides key insights into the study of urban resources. Furthermore, associated engineering disciplines such as civil engineering are professional in nature and accreditation and do contribute to the development of urban metabolism. In this regard, the work of environmental resource management often bridges methodological and intellectual gaps between environmental scientists, engineers, planners, and industrial ecologists.

Industrial ecology as applied to the challenges of contemporary cities is treated in detail in chapters 4 and 5.

Urban and Regional Planning

The disciplines of the regional and urban planner are inherently oriented toward future visions of alternative urban scenarios. Recently, partly driven by sustained and concerted demands for guidance by municipal authorities, urban and regional planners have been formulating strategies for resource efficiency in the urban context. However, there is a lag between the uptake of new models and perspectives of the respiration of the urban space and the professional conventions of the planning field. This can be understood as a reaction against some elements of urban metabolism that may seem reductive to the planning world, which is routinely engaged in multiple facets of urban life including social, cultural, historic, architectural, and urban infrastructure design. While urban metabolism is decidedly focused and comparatively narrow in its range of concerns, planning delves into every facet of the life of cities.

However, the need to reduce urban resource consumption, and especially urban energy intensity, has come to be a central priority for cities around the world. For example, urban economic development is beginning to appreciate the value of supporting low-energy housing retrofits as it benefits local communities. Creating jobs and supporting emerging businesses raises tax revenue while alleviating the burden of escalating fuel prices for home heating and cooling. It also provides household economic relief and releases discretionary income to be spent in local businesses. As a result, a focus on urban resources is now a priority for many cities.

Evidence of this can be seen in the many green city plans developed within the last few years. Chicago; New York City; Boston; the cities of Portland in both Oregon and Maine; Austin, Texas; Boulder, Colorado; San Francisco; and Santa Monica, California, in the United States, as well as Zürich and Geneva, Switzerland; Lyon and Paris, France; Berlin, Munich, Hamburg, Dusseldorf, Köln, and many others in Germany; Lisbon, Oeiras, Oporto, and others in Portugal; Madrid, Spain; Singapore, and many others have invested the time and funds to develop environmental action plans and green city initiatives. They have hired consultants, tapped academic researchers, and developed their own "in-house" expertise to tackle the most pressing resource and environmental issues pertaining to their respective locations.

Both within their own planning departments and agencies, and with their partners in academia and private engineering and design firms, planners hold particular sway with municipal policy makers in many of these enlightened cities. The challenge remains one of a combination of the paucity of practical tools appropriate for use by planners, and the organizational and institutional legacies that often present obstacles for integrating policies across municipal systems and infrastructure. This is a gap that can be and is being filled by organizations that provide straightforward decision guides for the planning profession, such as the International Council for Local Environmental Initiatives' Sustainability Planning Toolkit and the United States Green Building Council's LEED (Leadership in Energy and Environmental Design) for Neighborhood Development Rating System.

Environmental Research and Engineering
Arising from the traditional engineering specializations, notably those of the civil engineer (water, transport, infrastructure, and buildings), environmental research and engineering combine knowledge, technology development, and implementation with systems analysis and modeling in addressing urban resource flows. The best indication that these professional disciplines have taken up the challenge of sustainable cities is the growth of substantial consulting services that are currently offered by some of the world's largest engineering firms. Well-known groups such as Arup, CH2M Hill, CDM Smith, and many other professional consulting outfits now offer engineering, planning, logistical analysis, and other services that are explicitly oriented toward providing practical solutions for cities interested in achieving gains toward resource efficiency, especially with regard to water and energy resources. These consultants have been key members of large-scale teams advising on the design and construction of new carbon-restricted or carbon-neutral cities, such as Dong-tan and Masdar. Private consultants are also important contributors to green city efforts, not only in providing consulting services but also in engaging with academic researchers and nongovernmental groups to advance knowledge and action (for example, Aggarwala et al. 2011).

While environmental engineering as applied to the design of water and waste systems has been able to incorporate the priorities of urban resource efficiency, this cannot be said of some other aspects of urban infrastructure. For example, the promise of the smart grid for urban energy efficiency, especially incorporating significant sources of renewable energy, is a less tractable problem. The extent of the electricity grid is regional, national, and international, and the relative lack of regional planning tools hampers a quick move toward an efficient and "green" power grid.

Tier Two

Architecture, urban ecology, urban economics, and economic geography and ecological economics constitute the second-tier group. The work of each these groups overlaps in significant ways with the prospects for a unified urban metabolism.

Architectural and Civil Engineering

Architectural professionals around the world are now deeply engaged in delivering on the promise of a new sustainably built environment. The mere facts surrounding the resource consumption of the built environment are stark reminders of the importance of attending to the efficiency of our buildings and infrastructure. In many developing countries, 30 to 40 percent of primary energy consumption is devoted to heating, cooling, and lighting buildings. That the construction and operation of buildings comprise one of the largest sources of urban resource demands places the architectural professional in the center of the effort to design and manage a sustainable built environment.

In addition, many organizations have identified the built environment as the most promising economic sector for delivery of cost-effective mitigation and adaptation strategies against the most dire consequences of global warming (for example, Intergovernmental Panel on Climate Change 2007).[1] The interest in green buildings has led to the emergence of a forest of organizations that provide guidance in designing and engineering material- and energy-efficient healthy buildings. Chief among these in the United States is the United States Green Building Council, which has now produced the LEED for Neighborhood Development rating system, one of the only guides specifically oriented toward architects and planners.

Urban Ecology

Urban ecology has been focused on urban ecosystems, with the intent to understand and eventually shape the interaction between socioeconomic urban activities and natural ecosystems (Collins et al. 2000; Grimm et al. 2008). Closely related to work in urban and regional planning that addresses the urban-nature interface, urban ecology derives its theories and methods from linking the science of natural ecosystems with sociological studies of urban systems and form (Pickett et al. 2001; Pickett, Cadenasso, and Grove 2005; Grimm et al. 2008).

Some of the most intriguing and potentially useful work to come out of this field are associations between urbanization and certain biogeochemical effects (Kaye et al. 2006; Clergeau, Jokimäki, and Snep 2006).

Urban Economics and Economic Geography

Understanding the economies of cities would seem to be an important aspect of economic theory and research. As drivers of national economies, centers of learning and innovation, and attractors of manufacturing and service firms, cities certainly deserve the attention of economists. Yet economists are latecomers to the study of cities, having been preceded by historians and sociologists. Urban economics and economic geography, the two subfields that address agglomeration economies, have only been developed during the second half of the twentieth century. Despite this, there has been a surge of interest and progress in urban economics. Some of its findings are useful to urban metabolism.

Why do people choose to locate in dense settlements? This is the core question for urban economics. The answers thus far are derived of a fundamental attribute of urban economies: that of spatial equilibrium. Cities attract people, firms, and construction in a seemingly mutually beneficial equilibrium contained within a relatively small spatial extent. Cities exist because of the dynamics behind this spatial equilibrium. It has been the work of urban economists and economic geographers to explain the mechanics of this equilibrium. The results of their work are intriguing for their relation to resource flows and intensity of urban production and consumption. Their work is also noteworthy for what it may tell us about the location choices of people and firms under new policy regimes meant to drive resource-efficient urban form and systems.

The governing attribute of urban spatial equilibrium, it turns out, is that a benefit in one location must be accompanied by an equilibrating cost in that same location. That is, urban economies lack the presence of arbitrage opportunities. This holds whether from the perspective of an individual or an organization, and leads to robust explanations for the presence of high wages and high rents in urban centers. This also holds true even when the benefit and cost are two different kinds of things—like good weather and high housing costs, or low rents in an underperforming local public school system (Glaeser 2008). All of these factors contribute to the assessment of utility in making location choices in the city for the individual (housing) or the organization (labor). That is, the utility of a location can be shown to be:

utility = income + amenities − housing costs − transportation costs.

The concept of spatial equilibrium has led to the development of models that attempt to illustrate the dynamic relationships between wages, transportation, and housing costs and amenities, all while balancing centrifugal and centripetal forces in the maintenance of the urban economy (Fujita, Krugman, and Venables 1999). Various models have been empirically verified to show that holding income and amenities constant leads to a close correlation between housing and transportation costs; similarly, holding transportation costs and amenities constant leads to a close correlation between income and housing costs (for example, the Alonso-Muth-Mills model; Alonso 1960; Mills 1967; Muth 1968). Incorporating the principal of spatial equilibrium operates at the level of explaining the origin of differences in urban-form concentration that serve location preferences of firms and households, and is a key goal of researchers working to modify transportation choices for improved time and energy efficiency, as is the relationship between the urban economy and the flow of global resources illustrated in figure 3.3.

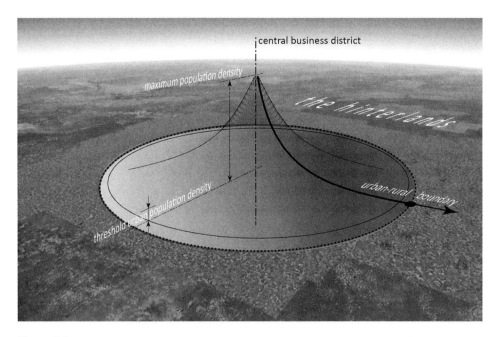

Figure 3.3
The idealized city as a centralized, population-density gradient.

The amenity element of these relations poses an interesting possibility for considering alternative urban scenarios. Amenities can include many kinds of sought-after elements of the urban context, such as proximity to a park, good schools, cultural facilities, access to recreation, and other spatially defined urban benefits. Negative amenities also play a role, such as a long commute, proximity to a waste facility, industrial zone, or power plant. For the most part, urban economics has developed the tools to associate these amenities with housing prices and income levels, and these associations have been shown to be good predictors of location preferences, urban morphology, and transportation choices.

Expanding the notion of amenities to include emerging alternative urban systems holds some analytical potential. A key intention behind the effort to incorporate principles of economic geography in the study of the resource consumption of cities is based in the assumption that learning something about agglomeration economies will better guide the theorist toward understanding changes in the nature of those economies (Fujita et al. 1999).[2] This is especially relevant to the urban metabolism project, as it expands the analytical toolkit available to assess the viability of alternative urban scenarios.

Today urban economics and economic geography have derived relations between people, firms, and construction activity that explain the city as a distinct economic entity governed by a special set of agglomerating forces (Glaeser 2008). While there is generally substantial discussion of the economic uncertainty of sustainable practices, little thought has focused on the relation of the principles of urban economics with the most promising sustainable practices in the urban context. In fact, while the economic perspectives formulated thus far go a long way toward explaining fundamental characteristics of the growth and maintenance of contemporary cities, the most pressing question is: how can this progress in urban economic theory assist the effort to develop alternative urban resource-flow scenarios?

Consider relating the fundamental framework of spatial urban economic equilibrium with the leading alternative resource-efficient urban scenarios. For example, perhaps building integrated and community-owned renewable energy production and storage within a city is considered an amenity by the residents of that urban district. In urban economic terms, this would correlate directly with housing, transportation costs, and income levels for that district.

Or consider the now commonly proposed alternative transportation scenarios for city-center personal transport linked to the electrical grid. Developing the infrastructure for such a system requires significant technical skills and management. It is clear, through research in urban economics, that another aspect of cities important to the maintenance of their economies (as manifest by centripetal agglomeration forces) is the density of skill, education, and entrepreneurship that cities promote. This capability for promoting innovation establishes a good business and technology context for such an alternative transportation scenario.

It is clear that an economic understanding of cities is essential for grasping the potential for change in the future, especially the kind of change that affects wages, labor availability, and housing and land rents. Significant work is now trained on the relationship between urban development and environmental issues (for example, Glaeser and Kahn 2008). Of particular interest in recent work is the relationship between transportation costs and agglomeration economics (Pflüger 2004).

Ecological Economics

Ecological economics is also integral to the project of urban metabolism, though the challenge in the urban context is equal to the larger challenge for ecological economics: that is, broad acceptance and implementation in society and business. However, as the central framework for accounting for material and energy flows and ecosystem services that feed the economy, ecological economics may play a central role in assessing environmental and ecological value in light of alternative urban scenarios for the future.

Ecological economics proposes an alternative grounding for economic thought, that of the physical world and thermodynamic laws. This includes the relationship between urban economies and the ecosystem services that provide materials and energy and act to disperse wastes from urban respiration.

Four aspects of ecological economics are most relevant to the development of urban metabolism.

First, ecological economics questions the delivery of resources from hinterland to the city, especially as related to progress on conceptualizing the urban-natural interface and the consideration of ecosystem services that provide value to urban activities (Gutman 2007). Associated with this line of work are suggestions for new arrangements in the delivery of resources to the city space.

Second, perspectives from the application of input-output methods have resulted in the formulation of strategies for reusing, recycling, and generally lengthening the longevity of materials in the economy. Co-location of industries and firms is another strategy that can be shown, through the application of input-output methods, to decrease the overall resource intensity of industry.

Third, ecological economics has brought analytical rigor to the valuation of ecosystem services and, by extension, environmental impacts of urbanization. This work leads directly to the formulation of urban sustainability indicators that are relevant to both economic development and environmental impact priorities.

A fourth set of issues defining the potential and character of green urban employment has been less served by ecological economics, though it is of critical importance to cities and their aspirations for green futures: this is the topic of green jobs. The early literature from industrial ecology contains little mention of the effect on employment that "structural shifts" toward dematerialization, decarbonization, and a new "green" economy may have. The focus in the early years of the discipline of industrial ecology (circa mid-1980s) was the grand project of describing contemporary societal metabolism in physical terms. Later, calls for a linkage to the monetary economy were made, and only then did the research community begin to speculate on the ramifications that a new *physically controlled economy* would have on the economy as a whole (Koenig and Cantlon 2000; Brunner 2002; van den Bergh and Janssen 2004).

This is in contrast to the veritable cacophony regarding emerging employment opportunities that early advocates believe will accompany the new green economy. Diverse voices have emphasized the probability of a new kind of employment sector, in between manufacturing and service, that innovates, provides, and maintains systems while delivering services to feed a new sophisticated and high-tech sustainable economy. While the academic community focused on the physical, scientific basis for this new reality, the nonacademic world has rather fiercely seized on employment as a major benefit of the cleaner, more efficient, technologically laden future.

This is understandable. On the one hand, employment is considered the standard bearer for economic health at every level and scale of the economy, from the urban district and neighborhood to the national arena and the global stage. Employment is particularly critical to the urban context because of the corollary effects that changes in the level of employment bring to cities. High unemployment is widely considered a direct cause of crimes of all types, including the most violent offenses. High unemployment has even been linked with increases in the incidence of mental illness, including depression and suicide.

Therefore, cities have been particularly keen to reap the potential rewards of a reorganization of the labor market due to the new, more sustainable economy. Green job creation is a cornerstone of this perceived opportunity. Almost every municipal initiative tied to renewable energy, green buildings, alternative transport, or other evolving sector of the new economy has cited the probable positive yield in new, better jobs (see Chicago's Climate Action Plan, for example).

Tier Three

Tier three includes two areas of study that attempt to explain the growth and behavior of contemporary urban form, and one that focuses on the complex social and cultural factors that are relevant to an assessment of future urban sustainability.

Urban Morphology, Scaling Laws, and Rank Size

The relationship between urban form and resource consumption has catalyzed interesting work in the mathematics of cities. One surprising characteristic of the mathematics of cities is the enduring quality of the interest in describing urban centers through scaling phenomenon (Strogatz 2009). While some have utterly dismissed scaling and power laws as nothing but statistical tautologies, the interest perseveres and continues to spawn work from diverse researchers (Decker, Kerkhoff, and Moses 2007). Much of this was initiated by the work of George Kingsley Zipf, in stating his now-infamous scaling law, Zipf's law.

Another set of key insights regarding cities is work that addresses the particular aspects and sometimes uncanny behavior related to their size and growth (Batty 2007; Bettencourt et al. 2007).

It is not yet clear how this work will be able to practically inform the effort toward resource-efficient urban metabolism. The prospects are intriguing but, as of yet, indistinct.

System Dynamics

Not a field per se, but a general methodology, urban system dynamics is worth mentioning here, since it has attracted researchers from a variety of fields and professions. As a focus of study, the contemporary city presents a particularly challenging subject for system dynamics. It is clear that system dynamics has a key role to play in describing the relationship between socioeconomic development, natural

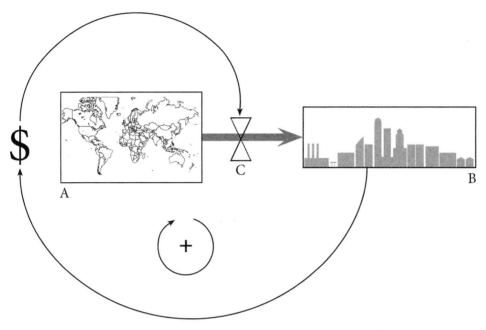

Figure 3.4
System dynamics diagram representing the fundamental relationship between terrestrial resource extraction (A) and the addition to stock of the city (B). The rate at which physical resources are driven toward urban consumption (C) is dependent on the intensity of the financial economy. Furthermore, the magnitude of the urban stock positively and directly affects the intensity of the financial urban economy.

ecosystems, and resource flows (Mäler 2000). Describing system behavior in a holistic manner such that physical flows and economic value are linked is a key goal of some of this work; see figure 3.4. Chapter 8 is devoted to the urban systems perspective, so this section will remain brief to avoid redundancy.

Sociology
Sociological studies indicate that the future of urban spaces depends to some extent on serious investigation of the demographic, social, cultural, and even psychological factors that are to be found in urban contexts. It seems obvious to state that demographic dynamics can serve to reveal the motivations and pressures that drive urban populations to change over time. Aging societies, rates of birth and death, and shifting composition of households, among many other factors, certainly affect the priorities and decisions that determine migration to and from cities. Similarly, cultural frameworks and shifting values, especially from one generation to the next, may influence the changing nature of the inner city and peripheral development, as well as modes of urban work and transportation choices. All of these factors

contribute to an urban psychology, variously represented as modes of urban thinking and intellectual production.

In the context of urban metabolism and urban resources generally, sociology maintains the position of steward of the nature and quality of urban living and working. Issues regarding the physical and mental health of the urban population as supported by spaces, institutions, norms, infrastructure, and other physical and non-physical urban systems has been the province of urban sociology for several decades (for example, Freeman 1987). Changing urban demographics, industrial and commercial mixes, and land use and development, among many other dynamic forces, are assessed in terms of criteria that articulate the livability of cities. Resource flows as affected by emerging trends such as urban farming, shared sustainable mobility, high-density neighborhood planning, and architectural design are discussed in terms of the framework of sociological theory.

Integrating Urban Metabolism

The prospects for breakthroughs in providing real pathways toward a sustainable society seem to be better positioned in a context of intensely complementary work in a variety of fields. These prospects are also made to seem more likely by explicitly linking the work in some key fields that, for one reason or another, may be converging toward each other. The three tiers outlined above imply both kinds of working relationships: complementary but not necessarily coordinated work, and directly linked efforts (see figure 3.5).

A lingering question remains in light of the integrated framework presented here: how can we integrate the near-certainty of resource constraints and environmental limits into a series of coordinated actions? Is it the province of the integrated framework, and all those involved, to assume roles in this other kind of integration?

It is in the lack of truly unequivocal efforts toward a sustainable society, based first on accepting the likelihood of certain sobering future realities and then acting on them, that continues to indicate coordinated and effective societal action is the real challenge. This is in light of well-founded research indicating a coming crisis on several fronts and holistic and broad calls for action and plans by many (for example, Odum and Odum 2006).

Despite these sobering thoughts, the breadth of fields involved in contributing to a future of resource-efficient and sustainable cities is a hopeful sign that there is great expertise and commitment to understanding every aspect of the behavior of cities. Coordinating and marshaling these fields toward coordinated action will require an explicit effort to codify processes, reach agreement on indicators, and communicate knowledge and experience of current initiatives from around the world.

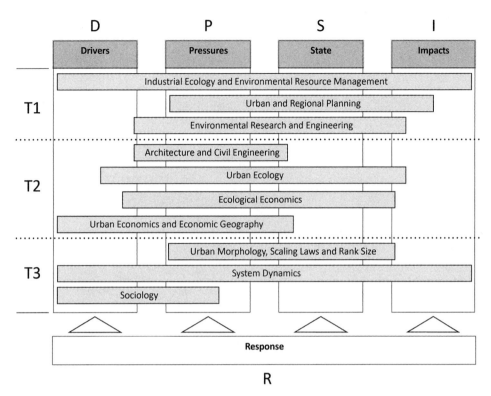

Figure 3.5
DPSIR framework showing the coverage of each field included in tiers one, two, and three.

II
Industrial Ecology: A Framework

4

Industrial Ecology: A Framework of Tools and Practices

The industrial ecology toolbox is aimed at providing a framework of methods and metrics to address the variety of dimensions and length scales required to understand urban metabolism. These range from micro to macro in three major dimensions: socioeconomic, environmental, and time. This framework will enable a better understanding of the contributions, limitations, and complementarities of the main tools and metrics to be analyzed. These include:

- Economic input-output methods *EIO methods are based in economic input-output tables that describe trade—sale and purchase relationships—between different economic sectors in a given country or region. The EIO methods were developed by Wassily Leontief (1906–1999), who was awarded the Nobel prize in economics in 1973 for this work. The method can be used to forecast and plan the production of economic sectors required to fulfill a given demand for different goods or services, and therefore has been widely used throughout the world in economic policy making. EIO tables are available for most of the world's economies, and this makes the method very useful. The method can be extended to perform environmental analysis.*

- Material flow analysis *MFA methodology is aimed at characterizing the material flows that enter, accumulate within, and leave a given system during a period of time. The boundary of the model is the socioeconomic system of one region, and therefore the focus can be set at the interaction between an urban area, a region, or a country, with the environment that surrounds it. The material flow indicators are physical system descriptors in which the material inflows of an economy are shown to be one measure of the economy dependency in resources and environmental burdens associated to their extraction, and the material outflows of an economy are one measure of the effective useful materials delivered and the losses that result in different forms of pollution. This exercise contributes to quantify the potential for improvement when an industrial ecology framework is adopted.*

- Ecological footprint *EF is one of the MFA tools that quantifies the amount of biologically productive land and sea area necessary to provide the resources a population consumes and requires to absorb its wastes, given current technology and management practices. This provides a unique benchmark that can be compared with the absolute maximum of total productive areas available on Earth, and constitutes a powerful argument to promote public perception of environmental problems.*

- Life-cycle assessment *A methodology for gathering information to evaluate the environmental impacts produced by a product or service over its entire life cycle. In the LCA methodology, product life cycle represents the extraction and production of raw materials and energy, manufacturing, marketing and distribution, use, and finally processing the product at the end of life. Economic input-output extensions that enable LCA are also discussed.*

This chapter concludes with a discussion of how the information made available by this framework of tools and metrics can support policy making.

Systems, Scales, and Tools

The scope and relevance of environmentally oriented urban policies have increased substantially during recent decades and, most importantly, their emphasis has expanded from concern about local, acute pollution problems, toward major, long-term regional and global issues, and from purely environmental concerns to a commitment to sustainable development. This requires an integration of economic, social, and environmental issues in a coherent framework, safeguarding the essential interests of each dimension as well as the extension of the focus to include local and distant regions and present and future generations, and thus calls for different scales to be considered.

In order to foster our capacity to make adequate collective decisions in this complex and dynamic context, and particularly in an urban perspective, we argue for a new intellectual framework, industrial ecology, and a systems approach that brings the analytical capacity to deal with the multidimensional, multiscale problems that are characteristic of our current challenges.

In this chapter we offers a toolbox that combines several methods and metrics to provide a significant contribution in understanding the complex relations between the drivers, pressures, and consequences of urban systems and their aggregate effects at regional and global levels, thus aiding policy makers in making more informed decisions.

This framework can be represented in a tridimensional diagram that includes the three main dimensions of the system under analysis, namely (1) environment, (2) socioeconomics, and (3) time, as represented in figure 4.1.

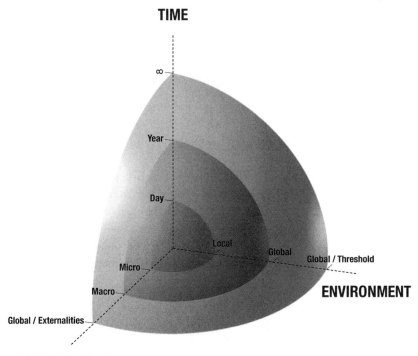

Figure 4.1
Tridimensional space of the metrics adopted to represent a complex system.

It shows that under each one of these dimensions, we may have multiple scales to characterize the system, starting with the possibility of a particular tool being independent of a given dimension, which we represent as "0," and then evolve in different scales, which were considered as follows.

• Environmental scales range from local to global, and to a special category which is represented as "global (threshold)" and is used whenever there is a known absolute limit level for the use or consumption of a specific environmental resource, beyond which we run into sustainability problems. For example, we know the area of productive land available on earth, and this constitutes an absolute limit that cannot be overcome, or if we need more energy than the "renewable" energy which we get mostly from the sun, we would also have problems of unsustainability in energy use, and therefore these are known absolute frontiers.

• Time scales range from instantaneous (considered as any phenomena whose characteristic time scale is shorter than one day) to scales of a year, decades, or infinity, which correspond to global trends.

• Socioeconomic scales are represented by microeconomic activities, at the level of an industry or business and its work force; macroeconomic activities, usually occurring at a regional or country level; and global activities, here corresponding to the impacts of the economic activities at large, even if they are not currently properly accounted for (as for example, in the case of externalities).

This analysis is important to promote a better understanding of the information required to analyze a complex system, the type of tools required to deal with it, and the dimensions of the impact associated with the choices made to improve the system efficiency.

As an example, let us consider a hypothetical transportation system in a city that is based on old diesel engine vehicles that cause poor environmental conditions, namely by emitting excessive particulates and volatile organic compounds, influencing human health negatively as well as the housing prices in those particular neighborhoods, represented by A1 in figure 4.2. The use of vehicles also has an impact on a global scale through global warming. If we represent the perception that people

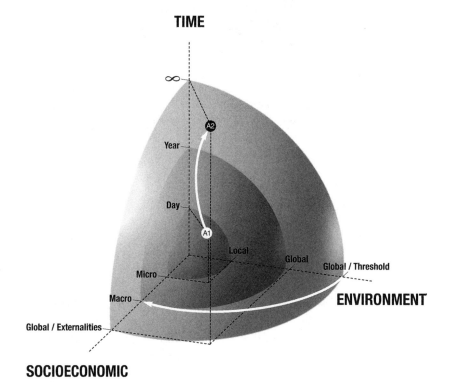

Figure 4.2
Problem representation in a tridimensional space.

have of this system in our framework, the environmental axis will be represented by a local (A1) and a global scale (A2), time scales typically of a day, as the pollutant concentrations vary on an hourly basis (A1) and normally vanish during the night, as the traffic decreases, but also at the scale of infinity, to represent the global effect of increasing CO_2 concentrations (A2). At the socioeconomic scale, this is mainly a microeconomic activity, as traffic results from people's need to be transported to their daily activities.

The analysis of this problem requires a set of tools that can address equivalent ranges of scales and dimensions of the system. In a DPSIR framework, it is also critical to understand if the field of solutions that policy makers may consider (responses) might push the problem to other dimensions and scales (not relevant before their intervention), thus requiring a broader set of explanatory tools than those anticipated by the nature of the problem.

This can be illustrated by considering a solution to the urban pollution problem mentioned previously, which consists in limiting traffic through the city center to electric vehicles. This solution would promote the use of electric vehicles, but consider if that electricity is produced by an old coal power plant characterized by significant emissions of global warming gases, particles, and sulfur. This new system will now be characterized by a global scale in the environmental axis. At the socioeconomic scale, this is mainly a macroeconomic effect, as well as being characterized at a global level by externalities that are associated with the environmental burdens caused by eventual acid rains due to sulfur emissions and global warming effects. The time scales are basically of several decades, represented by ∞.

As a consequence, the implementation of this solution is an example of how we can shift burdens from a set of environmental and socioeconomic impacts with different scales to another, in this case solving a local problem but not necessarily contributing to reducing global problems. This problem can be adequately perceived in the suggested framework, as represented in figure 4.3, and the full perception of the space of the problems/solutions as well as of the tools, constitutes the basis for a more thoughtful analysis that considers the problem and the solution spaces in all their dimensions and scales.

This analysis, and its representation in diagrams such as those represented in figure 4.2, is not only useful to provide a representation of the dynamics of the pressures and responses in a complex system, but also to support the selection of a set of tools and metrics adequate to providing a detailed understanding of these complex interactions, such as, for example, the symbiosis between the city and the global level. Providing a perspective on available tools and framing them under the tridimensional framework offered constitutes the main objective of the following section.

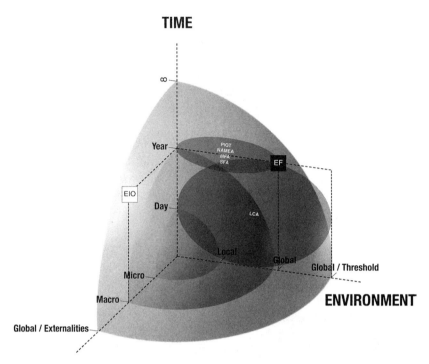

TIME

ENVIRONMENT

SOCIOECONOMIC

Figure 4.3
Tridimensional representation of the space covered by PIOT, NAMEA, MFA, SFA, LCA, and EIO.

A Framework of Tools

The need for an industrial ecology toolbox is based in the assumption that the characterization of the physical nature of human economy is vital for understanding the sources and full nature of the impact of society upon the natural environment. It is similarly assumed that effective strategies toward sustainable development will rely on the understanding of the physical phenomena associated with material and energy flows and their environmental and economic consequences in the appropriate scales and times.

In this context, it is critical to establish a toolbox intended to respond proactively to sustainability challenges. The new paradigm requires a change in focus, from the output side of the production system (which has been the foundation for environmental policies in previous decades) to a complete understanding of the physical dimension of the economy, in order to enable a vision of the economy as a set of processes that extract materials from nature, transform them, keeps them as society's stocks for a certain amount of time, and, at the end of the production-consumption chain, disposes of them again in nature. This is critical, as it is widely recognized that environmental problems arise at every step in this complex chain, which should therefore be addressed as upstream as possible.

Such a comprehensive toolbox can make use of available economic data that characterize the complex interactions between economic sectors, but requires new approaches to environmental accounting, such as the promotion of material flow accounting, which focuses on the "physical economy" in a comprehensive and integrative manner. MFA refers to accounts in physical units (usually tons) comprising the extraction, production, transformation, consumption, recycling, and disposal of materials (substances, raw materials, base materials, products, wastes, and emissions to air or water).

MFA has been supported by the need to further understand the physical dimension of sustainability, which is defined by the European Environmental Agency (1999) as the capacity to leave intact—for an infinite length of time—the stability of the internal evolutionary processes of the ecosphere, a dynamic and self-organizing structure. This is to say that an economic system is environmentally sustainable only as long as its supporting systems are physically in a (dynamic) steady state; i.e., the amount of resources used to generate welfare are permanently restricted to a size and a quality that does not overexploit the sources, or overburden the sinks, provided by the ecosphere. Progress toward sustainability will, then, depend at least in part on the reduction of resource input on the one hand and the further reduction of pollutant output on the other hand. Since material flows through the anthroposphere connect both of those ends, MFA offers considerable potential to help assess the path toward sustainability (Bringezu 1999).

The ecological footprint is a particular element of the MFA toolbox, which, although it is a simple indicator that understands material flows in terms of the amount of productive land required to provide them, benefits from this simplicity by being able to convey a strong and widely accepted message regarding the risk in exceeding the carrying capacity of the Earth. Accordingly, it is highlighted here.

The next logical step is LCA in that it provides an environmental extension of the materials-based analysis by correlating the material and energy flows required along a full product life cycle with their multiple environmental impacts. This effort has been leading to new techniques that combine economic tools, MFA tools, and LCA tools, which will be analyzed in this chapter.

These tools vary in terms of the dimensions covered: spatial, socioeconomic, or time; and in the degree of detail with which they resolve the system parameters, i.e., the characteristic scales of the phenomena they can model and represent. The next sections focus on each of these tools and explore their role in the framework depicted in figure 4.1. The objective is to have a toolbox able to model a system (for example, an urban area), at the same dimensions and scales as those that are characteristic of the main phenomena that affect it, in order to enable an adequate understanding and a framework that proves effective for supporting the policy-making process.

The main characteristics of the methods selected to be included in the industrial ecology toolbox, namely economic Input-output tables, material flow analysis, ecological footprint calculations, life-cycle assessment, and hybrid tools are discussed below.

Economic Input-Output Tables

Economic input-output tables describe the interdependence among sectors of a given economy, making use of a set of linear equations expressing the balances between the total input and the aggregate output of each commodity and service produced and used in the course of one or several periods of time (Leontief 1986).

Thus, considering the economy divided into n sectors, and if we denote by X_i the total output (production) of sector i and by Y_i the final demand for sector i's production, we have

$$X_i = z_{i1} + z_{i2} + ... + z_{ij} + z_{in} + Y_i \tag{4.1}$$

for $i = 1, ... , n$ and $j = 1, ... , n$,

where the z terms on the right-hand side of equation (4.1) represent the inter-industry sales by sector i to sector j. Thus, the entire right-hand side represents all of the inter-industry sales, z_{ij}, and Y_i the demand of sector i. Hence the sum over j represents the total output of sector i.

In order to make economies with different dimensions comparable, it is useful to divide the inter-industry flows from i to j by the total output of sector j, which is quantified by a technical coefficient, a_{ij}:

$$a_{ij} = \frac{z_{ij}}{X_j}. \tag{4.2}$$

The a_{ij}'s are fixed relationships between a sector's output and its inputs, and constitute the technical coefficients matrix A (A_{ij}). An explicit definition of a linear relationship between input and output is therefore assumed. Equation (4.1) can thus be rewritten as:

$$X_i = A_{ij}X_j + Y_i \tag{4.3}$$

The output required from each sectors to satisfy an increase in demand Y, can be quantified by

$$X = (I - A)^{-1} Y, \tag{4.4}$$

where $(I - A)^{-1}$ is commonly referred to as the *Leontief inverse*. A detailed derivation of the input-output methodology is provided by Miller and Blair (1985) and Leontief (1986).

Equation (4.4) can be reformulated as

$$X = (I - A)^{-1} \cdot Y = Y + AY + A^2 Y + A^3 Y + \ldots + A^\infty Y, \tag{4.5}$$

where the components associated with the direct contributions from different sectors to fulfill the demand, Y, are

$$X_{Direct} = Y + AY \tag{4.6}$$

and the indirect contributions, which are also referred to as of second and higher order, are:

$$X_{Indirect} = A^2 Y + A^3 Y + \ldots A^\infty Y. \tag{4.7}$$

If we want to establish an analogy with life-cycle assessment, it can be suggested that the indirect contributions that account for second and higher orders correspond to the upstream processes of the inventory associated with a product life cycle.

Material Flow Analysis

MFA is acknowledged as a set, or family, of models that focus upon a given geographical area, characterized by the systematic physical measurement of the magnitude and "location" of the mass of specific flows of environmentally significant materials for purposes of environmental monitoring, analysis, and management, as discussed by Niza and Ferrão (2006). It examines national or regional economy-wide driving forces behind induced flows that incorporate most major materials metabolized in the economy, though at various levels of desegregation—in terms of identifying flows between economic system components (e.g., sectors) and between these components and the natural environment.

In the context of the definition adopted for the MFA tools, we can identify four main approaches: (1) the bulk MFA, which emphasizes the bulk material flows in the system; (2) the physical input-output tables, where a greater level of detail is provided, (3) the substance flow analysis, which quantifies the flows of a given substance across a region; and (4) the ecological footprint as a carrying capacity–related method.

Bulk MFA

One of the most frequently used methodologies for performing MFA at the national level is that adopted by Eurostat, the statistical office of the European Union, which considers each country's economic system included in the domestic environmental system, and which, in turn, is included in the global environmental system.

This method is particularly useful to derive relevant indicators that are critical to characterize the economy's metabolism. From the material input side, a main indicator can be derived, the direct material input (DMI): the total amount of materials extracted from the domestic environment or imported, as raw materials, intermediary, or final products, that enter the economic system. The main output indicator is the domestic processed output (DPO)—the sum of the "outputs to nature." This includes emissions to air and water, wastes deposited in landfills, and dissipative flows. Although material flow indicators are evaluated aggregating materials by mass, water and air are generally excluded because these flows are orders of magnitude higher than others and their use could hide the weight of other materials.

Eurostat (2001) established a methodological guide for MFA and derived indicators. It comprises overall material flow balances and aggregated indicators on total resource requirements, resource efficiency, and total domestic outflows to the environment. The guide drew from the experience of the Wuppertal Institute and several national statistical offices of E.U. member states, and can be expressed making use of table 4.1, which summarizes the main indicators grouped by category.

It should be mentioned that the accounting rules consider some balancing items, which normally have to be included in the combustion and respiration processes, namely:

- on the input side, oxygen for the combustion of fuels and for the respiration of humans and livestock.

- on the output side, water vapor and CO_2 from the combustion of fuels and the respiration of humans and livestock.

A brief description of the indicators represented in table 4.1 is provided below.

- *Direct material input*   DMI is the flow of natural resources and commodities that enter the economy for further processing.

- *Hidden flows* are the movements of the unused materials associated with the extraction of raw materials, domestically and abroad: (1) materials that are extracted from the national environment but that are not used by the economy (such as mining overburden or soil excavation during construction); (2) the upstream resource requirements associated with imported products (such as the mining overburden arising abroad to make imported raw materials available; EUROSTAT 2001).

- *Total material requirement*   TMR indicates the total volume of throughput of the economy and therefore includes the potential amount of wastes and emissions. It provides an overall estimate for the potential environmental impact associated with natural resource extraction and use (Adriaanse et al. 1997).

Table 4.1
Overview of material flow indicators (based on Eurostat 2001)

Indicator category	Indicator		Accounting rules
	Acronym	Full name	
Input	DMI	Direct material input	DMI = Domestic raw materials + Imports
	TMR	Total material requirement	TMR = DMI + HF
	HF	Hidden flows	HF = hidden flows domestic + hidden flows from imports
Output	DPO	Domestic processed output	DPO = Emissions + Waste
	DMO	domestic material output	DMO = DPO + Exports
	TDO	Total domestic output	TDO = DPO + hidden flows domestic
	TMO	Total material output	TMO = TDO + Exports
Consumption	DMC	Domestic material consumption	DMC = DMI – Exports
	TMC	Total material consumption	TMC = TMR – Exports- hidden flows exported
Balance	NAS	Net addition to stock	NAS = DMI – DPO – Exports
	PTB	Physical trade balance	PTB = Imports – Exports

• *Domestic processed output*   DPO is the total weight of materials that have been used in the domestic economy and flow to the domestic environment. These flows occur at the processing, manufacturing, use and final disposal of the economic production-consumption chain.

• *Total domestic output to nature*   TDO, by including the domestic hidden flows, represents the total quantity of material outputs to the domestic environment caused directly or indirectly by human economic activity (Matthews et al. 2000).

• *Domestic material consumption*   DMC is the total amount of material directly used in an economy. In other words, it is the annual material flow that will either remain in the domestic economy as a material addition to stock (net accumulation— NAS) or will be released to the environment as part of domestic processed output (DPO excluding oxygen). It is, thus, the direct material flow that actors at all levels of the economy—households, firms, municipalities, politicians, and so on—have to manage with respect to accumulation, recycling, or final disposal in the environment (Eurostat 2001).

• *Net additions to stock*   NAS represents the net additions to stock and describes the annual accumulation of materials within the economic system. It can be said that it measures the physical growth rate of an economy (buildings, other infrastructure or long-lived durable goods as cars, for instance).

• *Physical trade balance*   PTB is the physical trade balance surplus or deficit of an economy. It expresses whether resource imports from abroad exceed resource exports of a country or world region and to what extent domestic material consumption is based on domestic resource extraction or on imports from abroad.

These indicators provide the only estimates of changes in the structure of materials use of an economy that have been produced in a consistent and reliable way (Hoffrén, Luukkanen, and Kaivo-oja 2001). It is thus clear that, for the sake of transparency, the political decision-making process can be supported by the availability of a set of simple and directionally safe indicators, applicable to different policy scenarios and thus contributing to the comparison of their potential outcomes (European Environment Agency 1999). Material flow indicators are rightly able to fulfill this objective.

Physical Input-Output Tables

Physical input-output tables (PIOT) quantify the material flows that are exchanged between sectors in an economic system as well as the material exchange across the system frontiers, which is to say that they provide information similar to the economic input-output tables presented above, but expressed in physical units (usually in tons) instead of monetary (value) terms.

As discussed by Giljum and Hubacek (2009), the most wide-ranging extension of PIOTs compared to EIO tables is the inclusion of the environment as a source of raw materials on the input side and as a sink for residuals (solid waste, waste water, and air emissions) on the output side of the economy. Thus, the resource flows that have no economic value and are therefore omitted in EIO tables, are also integrated into PIOTs. For each sector, the sum of all physical inputs has to equal the sum of all outputs to other economic sectors and to final consumption (e.g., private households), plus waste and emissions disposed to the environment. Concerning the changes in fixed assets and the interrelations with the rest of the world, the accumulation of materials (net addition to stock) and the physical trade balance provides information on the net difference. By definition, physical accumulation plus physical trade surplus or deficits has to equal zero.

However, in PIOTs, to enable a material balancing on the sectoral level, one thus has to add waste and emissions arising from production. The material balance is then equal to

Total output = input of raw materials + total input of goods and services – waste and emissions,

all in physical terms.

It can also be valuable to explore the compatibility with existing economic dimensions and national accounting methods, and therefore PIOT offers an excellent opportunity for integration or economic calculations using sectoral monetary transactions and values in national accounts in order to characterize the economy metabolism.

PIOTs can provide a very relevant framework for adding detail to a bulk MFA analysis, particularly at an urban level, where it can be used to characterize the correlations between economic activities and their associated material flows, thus leading to very interesting cross-correlations and economic material intensity indicators which are critical to characterize the urban metabolism.

In an urban context, data needed to account for material flows of an urban system may include:

- *inputs* Domestic extraction of resources, imports of raw materials and products;
- *outputs* Emissions and wastes, exports of raw materials and products.

If an adequate methodology is adopted, these methodologies may be useful to derive the following variables associated to the urban metabolism:

- throughput of materials, per material category;
- material consumption of each economic sector;
- waste treatment per material category and treatment type.

It is clear that an analysis based on PIOTs may require multiple tables, and the models may benefit from an adequate framework of tools by integrating economic and material flows information in the different tables. For example, in an urban context, we may use different tables, such as:

- the materials table providing the input and output flows of materials of the different economic activities;
- the throughput table providing the materials that are added to the city materials stock;
- the waste treatment table characterizing the waste destinations according to different treatment technologies; and
- the activity sectors matrix intended to distribute materials consumption through different economic sectors.

The selection of an adequate, multidimensional and multiscale toolbox will depend on the specific metabolism of the economic systems to be analyzed.

Substance Flow Analysis

Substance flow analysis (SFA) is a method focused on following different flows of a substance, characterized by a given chemical composition, across a system, normally a geographic region. This substance is normally integrated in different products that cross the system and, therefore, this method requires a detailed analysis of the composition of all the products that flow in the system. In this sense, this method is a subset of a material flow analysis and requires much more detailed information. Typical examples in existing research include nutrients like nitrogen and phosphorous, but also major toxic pollutants such as chromium, mercury, lead, and other heavy metals.

In this context, SFAs are normally requested by local authorities when a major environmental pollution problem resulting from a specific (typically toxic) environmental pollutant source is detected. In this context, it is generally associated to local scales and normally has a relatively high time resolution.

Hence, as discussed by Daniel and Moore (2002b), while a complete SFA may consider system-wide aspects linked to the single substance (group) under study, it will only describe a very small portion of the surrounding metabolism of the wider economic system. On this basis, SFA is only a "partial macro-MFA," which is powerful in terms of its specific problem theme but is limited with respect to MFA's potential strength in providing system-wide understanding of the nature of society's metabolism.

An SFA provides the information basis for overall environmental management and the formulation and implementation of elimination, reduction, dissipation minimization, and recycling strategies for dealing with a specific pollutant or other resource substance flow, as discussed by Daniel and Moore (2002b). This is achieved by identifying (i) existing levels and trends in substance emissions, accumulations and concentrations in the anthroposphere and various environmental media (e.g., the lithosphere and biosphere); (ii) the contribution of specific life-cycle stages and processes, goods, sectors, and activities to anthropogenic flows and emissions (and how these contributions change over time); and (iii) the qualitative nature of environmental impacts of certain human economy activities associated with a specific substance. SFA is also useful for monitoring and prediction purposes, for evaluating management strategies and for revealing the location, timing, and extent of future problems linked with specific pollutants or other high-impact substances.

However, as pointed out by Daniel and Moore (2002b), the "problem" substance flow measurement procedure is built around the specification of the physical parameters of particular embodied substances, and it diverges considerably from the bulk MFA method. It is implemented by establishing the quantitative physical flows of input, output, import, and export "goods" (e.g., detergents, paper, food steel, waste water) between human processes (e.g., economic sectors of production and consumption, capital operation). The identification of actual substance fluxes is derived from the flow of "goods" using transfer coefficients and other known concentrations and chemical relationships or by carrying out more resource-intensive primary research. Detailed analysis of these embodied and dissociated substance flows are linked to the tracking of metabolic pathways to determine the magnitude of flows through, and accumulations in, different human economy and natural environment compartments (for example, in the latter, the atmosphere, water, and solid media).

Ecological Footprint

The ecological footprint (EF) is a method that quantifies the amount of bioproductive land and sea areas required to make available all the products and services that we need, if they were produced with naturally available goods, inherently renewable.

The biocapacity of any area of land or sea represents the biosphere's ability to meet human demand for material consumption and waste disposal. As discussed by Kitzes et al. (2008), the Ecological Footprint and biocapacity accounts cover six land use types: cropland, grazing land, fishing ground, forest land, built-up land and carbon uptake land (to accommodate the carbon footprint).

Although the method for calculation of the EF is beyond the scope of the present text and is available in Kitzes el al. (2008) it is relevant to mention that the productive area required to the produce a given amount of a product harvested or to absorb a waste emitted depends on the national average yield for the specific product to be harvested and on the type of land available in the specific case study, as its productivity depends on the productivity of that soil or sea area. The productivity is quantified making use of equivalence factors that translate the area supplied or demanded of a specific land use type (i.e., world average cropland, grazing land, forest land, fishing grounds, carbon uptake land, and built-up land) into global hectares, units of world average biologically productive area.

In the case of urban systems, it is worth mentioning that the equivalence factor adopted for built-up land is set equal to that for cropland, and carbon uptake land required to compensate the use of fossil fuels is set equal to that for forest land. This reflects the assumptions that infrastructure tends to be on or near productive agricultural land, and that carbon uptake occurs on forest land.

The equivalence factor for hydro area is set equal to one, which assumes that hydroelectric reservoirs flood world average land. The equivalence factor for marine area is calculated such that a single global hectare of pasture will produce an amount of calories of beef equal to the amount of calories of salmon that can be produced by a single global hectare of marine area. The equivalence factor for inland water is set equal to the equivalence factor for marine area.

As an example, Kitzes et al. (2008) show that in 2005, for example, cropland had an equivalence factor of 2.64, indicating that world-average cropland productivity was more than double the average productivity for all land combined. This same year, grazing land had an equivalence factor of 0.40, showing that it was, on average, 40 percent as productive as the world-average bioproductive hectare. Equivalence factors are calculated for every year, and are identical for every country in a given year.

The national footprint accounts calculate the ecological footprint and biocapacity of individual countries and of the world. According to the 2008 edition of the National Footprint Accounts, humanity demanded the resources and services of 1.31 planets in 2005. This situation, in which total demand for ecological goods and services exceeds the available supply, is known as overshoot. On the global scale, overshoot indicates that stocks of ecological capital are depleting or that waste is accumulating.

This characteristic of the method, which quantifies a carrying capacity, facilitates its general understanding, and allows for its representation at the global (threshold) scale in the environmental dimension depicted in figure 4.1, represents the framework adopted in this text. This method has a time resolution of one year, as normally it is quantified for a given year and is independent of the economic dimension, as for other MFA methods.

It is clear that a method that expresses all the material flows required for human activity in a single indicator, such as the EF, may have many weaknesses for quantifying the economy metabolism, but this weakness is also its strength, in that its simplicity and intuitiveness have provided a very powerful message to convey the magnitude of human impacts when we compare it with the Earth's carrying capacity, making use of the notion of absolute limit established for ecologically productive land types and the fact that we have already exceeded it. This has been perceived by many and promoted a significant number of the responses we observe nowadays in policy making.

Life-Cycle Assessment

Life-cycle assessment (LCA) is based on the concept of the product life cycle that is associated to the consecutive and interlinked stages of a product system, from raw material acquisition or generation from natural resources to final disposal, as defined in International Standards Organization 14040. In a way, this concept stems from the industrial ecology metaphor as it is transposed from living organisms to products. If living organisms are born, reproduce, and eventually die, products are manufactured from raw materials, transported to the users where they execute their "function," and disposed of, to be eventually recycled or reused and close the cycle, or incinerated or landfilled.

In this context, the product's life cycle can be broken down into different stages, such as:

1. raw material extraction and processing
2. product manufacturing
3. packaging and distribution to the consumer
4. product use and maintenance;
5. end-of-life management: reuse, recycling, or disposal.

In every stage of their life cycle, products interact with other systems through the exchange of material and energy flows, and it is in this context that LCA is based on MFA methods. However, LCA extends this context in that the data-collection process leading to the preparation of an inventory of material and energy inputs and outputs for each life-cycle stage of the product is followed by an "impact assessment" that involves a conversion of the list of the various physical flow magnitudes, to the classification and quantification of the magnitude and significance of the environmental impact relevant to the natural environmental and to humans.

Life-cycle assessment aims at specifying the environmental consequences of products or services from cradle to grave, and in ISO 14040, LCA is defined as the "compilation and evaluation of the inputs, outputs, and potential environmental impacts of a product system throughout its life cycle." The core application of LCA consists on promoting product-related decisions support, as it can be used in product comparison, ecodesign, or to promote market claims.

According to ISO 14040, LCA includes four methodological steps:

1. goal and scope definition
2. life-cycle inventory analysis
3. life-cycle impact assessment
4. life-cycle interpretation.

In the first step of an LCA, the goal and scope of the study should be determined. This includes the precise formulation of what is to be analyzed and the definition of system boundaries.

The life-cycle inventory analysis specifies the processes that constitute the life cycle of a product and provides detailed information on all the inputs and outputs of material and energy that are associated with each of these processes.

The inventory table is used in the third step, the life-cycle impact assessment. In the impact assessment, the results of the inventory analysis are interpreted in terms of the impact they have on the environment. First of all, they are classified according to the kind of environmental problem to which they contribute. Examples of environmental impact categories include acidification, global warming, and human toxicity. Naturally, each emission can contribute to several types of problems. In the next step, the characterization step, the contributions to each environmental problem are quantified. For this, equivalency factors are used, which indicate how much a substance contributes to a problem compared to a reference substance.

The last step of the impact assessment involves the overall comparison of the environmental problems. If desired, the LCA study can be concluded with a single figure, an eco-indicator, in which each environmental problem is weighted in terms of its importance. This eco-indicator allows an easy and direct comparison of different products or options, but the weights used are subjective, and as a consequence these results should not be used for public assessment of the environmental performance of a product or for its comparison with any other.

The analysis of the results may also be used to promote the improvement analysis stage, which is intended to use the impact information obtained to reduce the environmental impact of the product or service by improvement product design (ecodesign) and/or processes and technologies used.

Hybrid Tools, Bridging Dimensions, and Scales

The major purpose of the industrial ecology toolbox consists in providing an adequate framework for modeling the physical nature of human economy as a vital step for understanding the sources and full nature of impacts of society and economy upon the natural environment. In this, it is assumed that effective strategies (responses) toward sustainable development will rely on the systematic collection of physical measures of material and energy flows.

This correlation between physical economy accounts and integrated accounts is convenient, since both try to overcome the inadequacy or incompleteness of monetary measures to describe the relationship between the human economy and its habitat. However, this requires a new set of tools that bridge the different dimensions analyzed, namely environment, economy, and time. These are the "hybrid tools" that result from the extension of those previously described and are analyzed in the next paragraphs.

Environmental Extension of EIO Tables

Integrating environmental and economic data has been a major effort of different researchers and organizations, and we highlight the United Nations method, the system of integrated environmental and economic accounting (SEEA). The SEEA-2003 comprises four categories of accounts:

- *Flow accounts for pollution, energy, and materials* These accounts provide information at the industry level about the use of energy and materials as inputs to production and the generation of pollutants and solid waste.

- *Environmental protection and resource management expenditure accounts* These accounts identify expenditures incurred by industry, government, and households to protect the environment or to manage natural resources. They take elements of the existing national accounting systems (NAS) that are relevant to the good management of the environment and show how the environment-related transactions can be made more explicit.

• *Natural resource asset accounts*   These accounts record stocks and changes in stocks of natural resources such as land, fish, forests, water, and minerals.

• *Valuation of non-market flows and environmentally adjusted aggregates*   This component presents non-market valuation techniques and their applicability in answering specific policy questions.

In short, the United Nation's SEEA attempts to integrate environmental and economic accounting by quantifying the services provided by natural capital as well as those of human-made capital. It covers the use or depletion of natural resources in production and consumption, and negative and positive changes in environmental quality from human intervention and natural regeneration or loss (United Nations 2000). It is an accounting tool that mainly follows the theories of internalization of environmental costs and that may be combined with other kind of data, such as physical data. In fact, if fully completed, the SEEA includes a monetary input-output matrix and also an MFA. This obviously requires huge quantities of data and, as a consequence, these accounts are almost uniquely held by statistics offices.

Another very important tool in this context is the National Accounting Matrix including Environmental Accounts (NAMEA), which provides an extension of monetary accounts by including the input of natural resources and ecosystem inputs, and residual outputs in physical terms per unit of economic value produced in the different sectors modeled in the national accounting systems presented in the EIO tables.

NAMEA relates environmental externality effects, measured in physical terms, to the monetary statistics in the national income accounting matrix classifications. These environmental impacts are typically aggregated into a limited number of topics to link ecological, social, and economic parameters and make national and sectoral comparisons in terms of contribution to total environmental pressures and sectoral environment intensities (e.g., CO_2 emissions per dollar of value added).

EIO-LCA: Economic Input-Output Life-Cycle Assessment

The EIO-LCA methodology complements the economic input-output analysis by linking economic data with resource use (such as energy, ore, and fertilizer consumption) and/or environmental impact categories (such as greenhouse gases emissions). At a European level, NAMEA can be used to account for greenhouse gas emissions, as they are represented in the form of a matrix (b) of gaseous emissions per economic sector.

Considering that B represents the vector of different greenhouse gas emissions (CO_2, CH_4, etc.), if b is a matrix of gas emissions per monetary unit of each sector's output, environmental impacts can be estimated by

$$B = b \cdot X = b \cdot (I - A)^{-1} \cdot Y . \tag{4.8}$$

As discussed by Suh (2004), a few different types of attempts to integrate the benefits of process-based analysis and input-output models have been performed, including the addition of input-output-based results to process-based models and desegregation of monetary input-output tables. A hybrid model that allows for full interaction between a process-based LCA model and an input-output model was suggested by Suh (2004), and constitutes the basis for the model discussed here.

In the hybrid method, a new algebraic formulation is adopted that includes in the same matrix the background processes associated with EIO data and the foreground processes that are specific to the system to be analyzed whenever we need to provide greater desegregation to the system under analysis. These processes are modeled, including material inputs, emission outputs, and their interaction with economic activity (the background system). Here, the foreground processes are those characteristic of the product life cycle under investigation, and the background processes correspond to economic sector activity, as represented in national accounting systems.

The integration of the two models must be done carefully, because on one side the foreground and background matrix have different units, and, on the other, it is necessary to avoid duplicating material/processes accounting.

The algebraic formulation of this model represents the external demand of process output in the foreground system, expressed as k. Considering that the technical coefficients of the foreground system quantify the products/commodities required in each process for accomplishing one unit of activity level, the technical coefficients are denoted by $\tilde{A}$, the technology matrix, where the use of tilde denotes any activity in the foreground system, and we have:

$$\tilde{A}.t = k . \tag{4.9}$$

This equation can be solved for t (the activity level required by each process) by inverting the technology matrix, $\tilde{A}$, and multiplying it by the vector of external demand of process output k.

$$t = \tilde{A}^{-1}.k . \tag{4.10}$$

Considering that the environmental burdens associated with the processes in the foreground system are expressed by $\tilde{b}$, the *intervention matrix*, since its coefficients represent interventions of the different economic processes in the environment: inputs (mainly extractions of resources) and outputs (mainly emissions of chemicals), the environmental interventions are expressed in the foreground process as

$$B = \tilde{b}.\tilde{A}^{-1}.k \tag{4.11}$$

If we consider the formulation of the emissions for the foreground system (4.11) with the one for input-output (8), the hybrid method can be represented by the following general expression:

$$B = \begin{bmatrix} \tilde{b} & b \end{bmatrix} \cdot \begin{bmatrix} \tilde{A} & M \\ L & I-A \end{bmatrix}^{-1} \cdot k \tag{4.12}$$

The methodology used to create the matrix of coefficients and to normalize the foreground and background units of the process can be calculated by the expressions (4.13) and (4.14), as discussed in Suh (2004). L and M denotes inputs from background and foreground systems to one another, respectively. In linking the foreground and background matrix, the dimension of elements for L and M matrices should meet with corresponding rows and columns. L shows monetary input to each sector per given operation time, while M shows total physical output per total production in monetary terms:

$$l_{pq} = q_{pq} p_p \tag{4.13}$$

$$m_{pq} = \frac{-a_{pq}}{p_p}, \tag{4.14}$$

where q_{pq}: input of sector p in each unit process q, p_p: unit price of product from sector p, and a_{pq}: technical coefficient from economic matrix. EIO-LCA thus has the capacity to provide engineering detail to the economic description of EIO tables, partially solving the problem associated with the aggregated structure of the economic input-output tables available, which represent economic sectors with different levels of detail, from country to country.

However, there are limitations in evaluating economic metabolism, such as the long time lags in data collection and table preparation, and related problems in discerning key patterns and trends in the vast array of generated information, which frequently precludes the availability of recent economic data.

This analysis shows that there are tools available to characterize economic metabolism that range from specific environmental analysis (LCA) characterized by local scales and microeconomics, to macroeconomic analysis (EIO-LCA), but it is clear that for intermediate levels of analysis relevant to sustainable product or systems design, we need alternative tools that bridge economics and environment with multiscale and multidisciplinary skills.

Life-Cycle Activity Analysis

Life-cycle activity analysis (LCAA) is a multidisciplinary tool that integrates engineering, environmental, and economical sciences, including operations research and solving for optimal solutions of multivariable complex systems, and can thus be interpreted as an alternative sustainable systems design tool.

Environmental analysis tools such as LCA or MFA, however valuable, generally do not include the description of economic mechanisms (allocation, optimization, substitution) or costs and benefits. Traditional economic models, on the other hand, have mainly focused on the general notion of externalities and do not explicitly describe the flows and transformation of materials.

In this context, an analytic tool, LCAA, was proposed by Freire et al. (2000) and Freire, Thore, and Ferrão (2001) to tie mathematical programming formulations of activity analysis to their environmental impacts. LCAA is based on the integration of activity analysis, a well-known procedure in economics, solving for optimal levels of production and for the optimal allocation of resources, with environmental LCA, which aims to quantify the environmental impacts of a product or a service from the "cradle" to the "grave."

The classical formulation of activity analysis distinguishes three classes of goods: primary goods (natural resources or materials), intermediate goods, and final goods (outputs). LCAA extends the concept of linear activities to embrace mass and energy fluxes over the entire life cycle of products. In particular, the proposed LCAA model includes one additional category of goods: "environmental goods," which represent the emissions of pollutants, energy consumption, and the dumping of waste. These environmental outputs can be further aggregated into a number of environmental impact categories, such as global warming, ozone depletion, and so on. This approach links up with the development of LCA methodology, and its aim is twofold. Firstly, it interprets the environmental burdens included in the output table in terms of environmental problems or hazards. Secondly, it aggregates the data for practical reasons, particularly for decision making.

The mathematical model of LCAA uses an input-output format, and may have the following formulation, as proposed by Freire et al. (2000, 2001):

Decision variables, to be determined

x is a column vector of levels of production activities,

t is a column vector of levels of transportation activities,

w is a column vector of supply levels of primary resources.

Parameters

Apr is a matrix of input coefficients; each element denotes the quantity of inputs required to operate a production activity at unit level

Atr is a matrix of input coefficients; each element denotes the quantity of resources (e.g., fuel) required to operate a transportation activity at unit level

Bpr is a matrix of output coefficients; each element is the quantity of outputs obtained when an activity is operated at unit level

Btr is a matrix of output coefficients; each element denotes the quantity of outputs emitted when a transportation activity is operated at unit level

cpr is a row vector of unit costs of operating the various production activities; it is known and given

ctr is a row vector of unit costs of operating the various transportation activities; it is known and given

crs is a row vector of unit costs of primary resources; it is known and given

d is a column vector of final demand; it is known and given

g is a column vector of environmental goals set by a policy maker.

The list of goods is partitioned into four classes: inputs of primary goods (P); intermediate goods (I); final goods (F); and environmental goods (E). Correspondingly, matrices A_{pr} and B_{pr} become partitioned into $A_{pr} = (-A^P, -A^I, 0, -A^E)$ and $B_{pr} = (0, B^I, B^F, B^E)$. Conventionally, one enters the *A*-coefficient of each input with a minus sign and the *B* coefficient of each output with a plus sign. This format includes the possibility of having $-A^E$, i.e., sinks of pollutants. Matrices A_{tr} and B_{tr}, however, are only partitioned into $A_{tr} = (-A_{tr}^P)$ and $B_{tr} = (B_{tr}^E)$, since the goods used in the transportation activities only include primary resources and environmental emissions (no intermediate or final goods are considered). The basic mathematical format of LCAA can now be written as the following linear program:

$$\min c_{pr} \cdot x + c_{tr} \cdot t + c_{rs} \cdot w \tag{4.15}$$

subject to:

$$-A^P_{pr} \cdot x - A^P_{tr} \cdot t + w \geq 0 \tag{4.16}$$

$$(-A^I_{pr} + B^I_{pr}) \cdot x = 0 \tag{4.17}$$

$$B^F_{pr} \cdot x \geq d \tag{4.18}$$

$$(-B^E_{pr} + A^E_{pr}) \cdot x - B^E_{tr} \cdot t \geq -g \tag{4.19}$$

$$x, t, w \geq 0. \tag{4.20}$$

To assure that for each intermediate commodity in each link there is conservation of the quantities of goods being produced, transported, and used in the subsequent activities, additional equations have to be included. In short, one equation is needed for balancing the quantity of each intermediate good leaving a region, and another equation should be added for balancing each intermediate good entering a region.

In addition, the *x*, *t* and *w* vectors may be bounded from above to reflect capacity constraints of production and transportation activities and on the availability of primary resources. Capacity bounds can also be included to reflect current behavioral patterns or to impose environmental policy options.

The objective is to minimize the sum of all current unit costs and the costs of all primary resources (4.15). Constraint (4.16) establishes the balance between the quantities of primary resources used by the activities and the amounts extracted from the environment. Constraint (4.17) states market clearing for the intermediate goods. Constraint (4.18) says that the demand must be satisfied. Constraint (4.19) states that the environmental impacts should be at most equal to the targets defined (vector g).

This formulation shows that LCAA integrates engineering, environmental, and economical sciences, including operations research, and that it solves for optimal solutions of multivariable complex systems and can be interpreted as a new industrial ecology tool, in that it can be used to promote optimum systems design, ecologically and environmentally.

The feasibility and potential of the LCAA methodology for optimizing the life cycle of products, with emphasis on alternative end-of-life processing activities, depends again on the availability of adequate data that might enable the analysis of integrated economic, environmental, energy, and product system models, developed and applied to specific case studies, but provides a relevant tool for systems design purposes.

The Toolbox: Schematic Representation of the Framework of Tools

The MFA tools vary in terms of the dimensions covered and their scales. They vary in whether they cover the national economy (bulk MFA), or specific economic activity sectors of a region as driving forces inducing material flows (e.g., PIOT), or whether they are designed to focus only upon the material flow and other environmental repercussions of specific goods, services, or processes, regardless of their location (e.g., SFA or LCA). This corresponds to different scales at the environmental dimension, and none of those consider any economic dimension, as all of them are expressed in terms of physical quantities and not in economic terms. Therefore, in our representation they are represented independently of the economic dimension.

They also vary in whether they are based on and compared with a virtual ceiling of nature's capacity to maintain its functions (carrying capacity methods), as the EF, which is thus characterized by a "global (threshold)" dimension at the environmental dimension and a macroeconomic scale.

Any of these MFA methods take place at a time scales of typically one year, and economy-wide accounts can then be held not only by bulk MFA but also by PIOT, EIO, NAMEA, and SEEA. However, when the purpose is to evaluate an economy's metabolism, the following limitations can be found.

- NAMEA and SEEA are integrated environment-economy accounts but require enormous data requirements. Consequently, there are long time lags in data collection and table preparation, and related problems in discerning key patterns and trends in the vast array of generated information. In the particular case of NAMEA, it doesn't account for physical stocks and that considerably diminishes it as a method for correctly evaluating economy metabolism.

- Even though PIOT and EIO-LCA can be used for the purpose, they are better suited for sector or activity fields' physical input-output description and not so much for economy-wide balances, due to the very restricted coverage of physical inputs and outputs compared with bulk MFAs.

On the other hand, MFA is acknowledged as a set, or family, of models that focus upon a given geographical area, characterized by the systematic physical measurement of the magnitude and "location" of the mass of specific flows of environmentally significant materials for purposes of environmental monitoring, analysis, and management. It examines national or regional economy-wide driving forces behind induced flows that incorporate most major materials metabolized in the economy, though at various levels of disaggregation—in terms of identifying flows between economic system components (e.g., sectors) and between these components and the natural environment. This makes them particularly suited to urban metabolism studies.

MFA supports a comprehensive physical metabolism analysis of an economy with the advantage that data requirements are much less than an input-output method, allowing year-by-year evaluations to perceive trends. It is then a technique with both capability of disaggregation and temporal response, allowing for the characterization of the dynamics of an economy's metabolism, as evidenced in subsequent chapters.

LCA constitutes another environmentally oriented tool, one used with the purpose of accounting for environmental impacts of products and services, along the complete life cycle, i.e., from cradle to grave. Considering that each industry is dependent, directly or indirectly, on a large set of other industries, this approach crosses different scales both in the environmental as well as in the time dimensions, though for MFA there is no economic dimension.

The representation of the scope of the MFA and LCA tools in the framework proposed is depicted in figure 4.3.

On the other hand, in the domain of the economic tools dedicated to explain the metabolism of an economy through the quantification of the monetary flows between different economic sectors, economic input-output tables have been established worldwide for decades and allow for accounting all the direct and indirect inputs to producing a product or service by using the input-output matrices of a national economy. Therefore, they include the macroeconomic dimension and yearly time scale.

This macroeconomic approach that characterizes the inter-industry effects of products and processes for a diverse set of commodities can be extended making use of environmental information associated with the emissions and other environmental burdens of each economic sector per unit value produced making use of the economic input-output–life-cycle assessment (EIO-LCA), as discussed before.

In addition, the bridging role of the hybrid EIO-LCA techniques, which enable the incorporation of any required details on particular processes inherent to the life cycle of the product to be modeled, contribute to overcome the limitation associated with the limited desegregation of the available economic IO tables, and offer an interesting compromise and integration between macroeconomic data and process-detailed information, thus resulting in a cost-effective strategy to promote product life-cycle assessments.

Finally, LCAA ties mathematical programming formulations of activity analysis to an activity's environmental impacts. LCAA is based on the integration of activity analysis, solving for optimal levels of production and for the optimal allocation of resources, with environmental LCA, which aims to quantify the lifetime environmental impacts of a product or a service. LCAA integrates engineering, environmental, and economic sciences, including operations research, as it solves for optimal solutions of multivariable complex systems, and can thus model systems in microeconomic scales as well as provides an environmental dimension that can range from the local to the global scales. This technique does also allow to model systems with different number of sectors and flows and allows for selecting optimal solutions among alternative scenarios.

The contribution of these tools in extending the range of application of the methods available is represented in figure 4.4.

Using the Industrial Ecology Toolbox in Decision Making

The exercise of highlighting the added value of industrial ecology methods in policy making is not new. The use of LCA tools as a basis for extended producer-responsibility policies or the advantages of MFA research for sustainable development policies, which have been highlighted in several reports (e.g., Adriaanse et al. 1997; European Environment Agency 1999; Matthews et al. 2000; Eurostat 2001), are increasingly accepted. These documents evidence that many of the environmental policies have thus far focused on wastes and pollution—on the back end of the materials cycle—even though more than half, and as much as three-fourths, of the natural resource use occurs at the front end of the process, before natural resource commodities enter the economy (Matthews et al. 2000). Since what leaves the

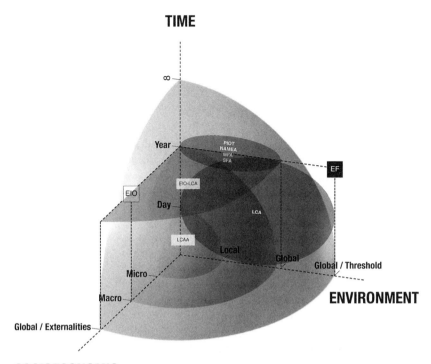

Figure 4.4
Additional domains covered by LCAA.

industrial system as wastes is closely related to the volume of material inputs, policies that reduce the use of base materials not only diminish extraction pressures,
but also wastes and pollution. Similarly, policies that make natural resource use
more efficient or increase recycling lower the requirements and the environmental
impact over the entire materials cycle.

In this context, MFA can be viewed as a tool that organizes and integrates
available primary data on materials extraction, product use, and final disposal. It
is therefore useful to provide flexible and quick responses to evaluate the effect of
increasing product demand as pressures to the extraction or release of specific
materials.

Matthews et al. (2000) express an interesting view regarding the importance of
MFA for policy. They state that it is not the purpose of accounting systems (like
MFA) to anticipate complicated issues by providing answers. Rather, they should
provide information that enables decision makers to ask the right questions. For
instance, concerning outputs (the subject of their report): which materials flow to

land, to water, and to air, and in what quantities? Which materials are dissipated into the environment with no or limited possibility of recovery? How much material is potentially recoverable and recyclable? And how are these outputs changing over time?

In policy terms, industrial ecology tools provide a variety of components that enable the combination of a pollution reduction perspective with a resource-minimization strategy. The industrial ecology toolbox approach allows not only localization of the predominant losses of substances to the environment but also accounting for the bulk material flows that are associated with their control in order to minimize resource requirements, as well as to quantify their broader environmental impact and their close interaction with economic activities, which normally constitute the main drivers for their circulation across economy.

Industrial ecology tools do also recognize that the number of "entry points" of material inputs into the economy is much lower than the number of "exit points," so whereas a manageable number of raw materials (maybe around a hundred) are extracted by only a few sectors (agriculture, forestry, mining and quarrying, construction), the material outflows and their related "exit points" seem to be innumerable (European Topic Centre on Waste and Material Flows 2003).

A large number of similar questions may be considered for the input and stock flows, and industrial ecology can help define the right questions. In order to organize this point of view, the contribution of industrial ecology to support decision-making processes may include the following features:

1. derive combined environmental/economic flow indicators, which may contribute to monitoring resource productivity and ecoefficiency

2. monitor the environmental pressure of the economy

3. information/awareness raising about environmental problems

4. design complex systems for optimal environmental performance at minimum cost

5. eco-design of products and services

6. promote environmentally and economically informed options between alternative products or services (that fulfill the same function)

7. measure the entire metabolic performance of the economy.

Metabolic studies are at the heart of industrial ecology and promote a better understanding how societies handle natural resources and their residence periods in the economy, and they are emerging in several ways at different levels.

Some studies clearly show that resource flows, such as construction minerals, are used by the economy for some decades; others, such as nutrition flows, may be released again to the environment within the same year. For example, around 80 percent of the total material input (DMI plus domestic hidden flows) into the German economy between 1991 and 1996 was released again to the environment within one year (Bringezu and Schütz 2001). A study covering the city of York concluded that 22 percent of all food brought into York is not eaten and ends up in the domestic waste stream (Barrett et al. 2002).

A study on material outflows from industrial economies (Matthews et al. 2000) evidenced (surprisingly, according to the authors) that the atmosphere is now the biggest dumping ground for the processed output flows of most industrialized economies.

In fact, evidence for the interest in MFA as a policy-making tool across OECD countries is that relevant studies and practical applications have progressed significantly. In some countries, this has even led to a move toward integrating MFA work in the national system of official statistics. Today, almost all OECD countries carry out some activity on resource and material flows and related indicators. It has become a widespread approach to environmental problems. As an example:

1. More than half of OECD member countries have developed or are developing economy-wide MFAs (Austria, Belgium, the Czech Republic, Denmark, Finland, Germany, Hungary, Italy, Japan, Korea, the Netherlands, Poland, Portugal, Slovak Republic, Spain, Sweden, Switzerland, the United Kingdom, and the United States).

2. In nine out of these nineteen countries, MFA work is a regular activity with annual updates.

3. In seven countries (Denmark, Hungary, the Netherlands, Poland, Portugal, Sweden, and the United States), MFA work has been carried out only on a stand-alone or pilot basis.

4. The input-output framework is used in Australia, Austria, Canada, the Czech Republic, Denmark, Finland, Germany, Italy, Japan, Sweden, and the United Kingdom in order to develop specific flow accounts distinguishing not only categories of materials but also branches of production. In some countries, hybrid flow accounts have been established by linking information from physical flow accounts to economic data from monetary input-output tables. In some countries, input-output analysis supplements work on economy-wide flows. Efforts are also being made for developing simplified PIOTs that could usefully be linked to economy-wide MFA.

5. In Europe, many countries have established NAMEAs, mainly in the field of air emissions and waste. Such accounts, and in particular their sectoral breakdown and their links to economic accounts, are often seen as useful complementary tools, and some countries have applied the NAMEA approach to material flow accounts.

6. In a number of OECD countries, research work has focused on studying flows of specific substances or groups of substances (e.g., Austria, Belgium, Finland, the Netherlands, Norway, Sweden, Switzerland, and the United States). Such work often concentrates on heavy metals and on other substances with potential negative impacts on the environment and human health.

7. Most OECD countries that have developed a national set of environmental or sustainable development indicators include in their set one or several indicators derived from natural resource or material flow accounting. In 13 OECD countries, MF indicators are part of proposed or agreed sets of environmental or sustainable development indicators (Austria, Belgium, Denmark, Finland, Germany, Hungary, Italy, Japan, Poland, Slovak Republic, Spain, Switzerland, United Kingdom).

In many OECD countries, goals and objectives concerning the efficient management and sustainable use of natural resources and materials have been embodied in national sustainable development strategies or environmental action plans but in only a few countries have time-bound quantitative targets been defined. Nevertheless, in general, these targets are not mandatory but rather an expression of desired policy directions. Particularly focusing on Europe, which evidence shows is the geographical region where material flows approaches are more developed and applied, the OECD survey found that material flows indicators are associated with broad policy goals in five countries. These goals are

- decoupling natural resource use from economic growth (Belgium: Federal Plan for Sustainable Development);

- using resources more efficiently (Denmark: National sustainable development strategy; associated indicators: e.g., TMR/capita, total consumption of selected resources)

- improving the efficiency of natural resource use and energy through the full life cycle (Finland: government program for sustainable development; associated indicators; TMR);

- achieving nontoxic and resource-efficient material cycles (Sweden: Swedish environmental quality goals);

- improving resource efficiency (United Kingdom: government framework for sustainable consumption and production; proposed associated indicators: decoupling indicators, among which are total material use and freshwater abstractions).

In short, industrial ecology tools are increasingly contributing to produce and organize relevant data in an integrated framework that provides an overview of the metabolism of an economy, understood as the structure and dynamics of the physical metabolism, and this is crucial to inform decision-making processes that may contribute to sustainable development, and in particular to urban systems.

5

Industrial Ecology as a Framework for a Sustainable Urban Metabolism

The industrial ecology metaphor is used to inform the formulation of sustainable urban governance strategies, through the definition of roadmaps designed as a set of sequential and complementary steps that are discussed in this chapter. This is combined with the DPSIR framework of indicators, which constitutes a unique tool for policy making, by providing a model that facilitates the understanding of complex interactions between drivers, their actions, and impacts and the responses that may improve urban sustainability.

An Industrial Ecology Vision for Urban Systems

The first urban systems were a creation of communities that became sedentary by dominating the techniques of farming and by domesticating animals. This occurred in locations selected for their natural support in providing basic goods and services, such as water, fertile soil, natural shelter, or favorable climate.

The development of sedentary settlements accelerated the population growth based on technological development, particularly making use of artifacts such as plows and agricultural processes, such as seed selection or irrigation. The technological development also resulted in better housing and food surpluses, and this allowed for social differentiation and economic specialization.

By 7,000 B.C. agricultural productivity supported the development of large communities numbering in the thousands. This also enabled a critical mass of people that promoted mutual protection and a division of labor enabling complementary functions to be developed such as farming, hunting, domestication of animals, health care, producing artifacts, or defense.

In this, the establishment of urban areas implicitly recognized that residents were embedded in natural ecosystems that had the capacity to renew the energy and materials required for survival and sustainability.

These regions that supported urban areas can be represented as "bioregions," and they are still today a major topic of debate, as discussed by Peter Berg (n.d.):

A bioregion is defined by the unique natural characteristics that occur throughout a particular geographic area, such as climate, landforms, watersheds, soils, native plants and animals, and other features. Every bioregion also includes human activities that should be carried out to join with these features in sustainable ways. Human inhabitation should be an interactive part of the ongoing life in a place. Bioregions differ greatly from each other, such as the contrasts between a coast on the ocean, a rain forest, an interior desert, or the Arctic Circle. Because of these vast differences bioregions require different ways for societies and individuals to relate to them in order for life to be sustainable. The way people live in New York, Beijing, Tokyo, or Berlin should reflect the wide bioregional variations that exist between the places where these cities are located. In order to plan, design, build, or direct human communities in ways that will achieve bioregional sustainability, we must consider the preservation of natural systems that are native to the place to be the basis for successful human inhabitation in it as well. People are ultimately dependent on the life of the place where they live. This isn't an incidental aspect of human life but instead must be adopted as a central and primary social fact.

In this context, bioregions have shaped the development of the initial urban settlements and are still very relevant, although in different ways. In fact, it can be argued that initial settlements have been dependent on their bioregions to prosper, as represented in figure 5.1, which characterizes the initial urban systems as including industrial or agricultural activities inside the "urban nucleus."

The more successful urban areas have developed a surplus of different goods and services, and this has induced social stratification and trade as a critical step for

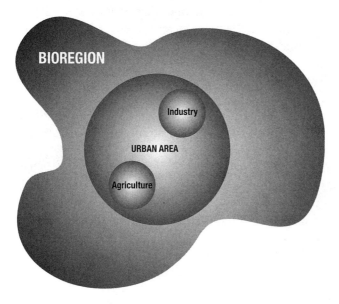

Figure 5.1
Urban systems embedded in a bioregion.

progress and economic development. The advent of commerce, together with the progress of transportation systems, leads to a globalized economy, where cities emerged as nodes of commercial activities that provide different goods and services and that contribute their unique role in continuing to attract an increasing number of people, providing enhanced opportunities for education, health services, technological innovation, and cultural development, making use of and agglomerating efficiencies resulting from the scale and density that characterize urban areas.

In this process, land price in the urban nucleus has risen, and increasing environmental concerns have promoted the decline of urban industries and urban agricultural activities, and these have been transferred to specific areas out of the urban limits, as represented in figure 5.2, leaving the urban nucleus to both cultural and higher-added-value activities, such as commerce and services.

The establishment of specific areas specialized in industrial and agricultural activities have resulted in a very complex system of global interaction, where products and artifacts used in a given urban nucleus, despite their local availability, may originate from very far away, depending essentially on their price and innovative characteristics, as represented in figure 5.3.

In this context, the sense of interaction between the urban area and its hinterland was broken and complex social structures emerged in large urban areas. Not only

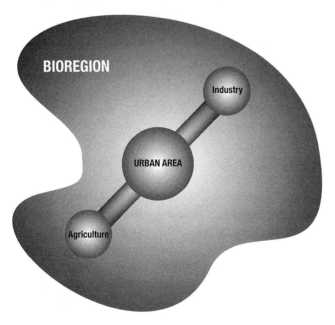

Figure 5.2
Evolution of the urban systems, the urban nucleus.

Figure 5.3
A global economy, beyond the bioregions.

the geographical ties were overcome, but others, such as family ties, eventually failed to provide the ambitious individual goals of social progression and were replaced by other bonds, such as ties based on economic gain or protection. Urban systems became interconnected systems, with a variety of nodes, such as: housing, markets, travel hubs, work, labor markets, networks, business supplier and consumer networks, energy systems, and water catchment and distribution systems.

The performance of the material and energy flows among those systems became crucial to their sustainability, particularly because their operation depends on more than locally available resources, and therefore, the user loses the sense of physical availability, which is increasingly replaced by economic wealth. Urban systems can be characterized by "inputs," such as raw materials and energy, and "outputs," such as products, goods, and greenhouse gas emissions. They also promote "stocks" inside the city, and these stocks change over time (such as intellectual capacity, social capital, or nutrients).

In these complex interactions, groups eventually developed different governing rules, hierarchies, beliefs, and values. The management of this complexity oriented for economic gain, encouraging the concentration of power, which in turn required large bureaucracies to administer the populations brought under the control of powerful individuals or groups.

This was reflected in the symbolic representation of that power, namely in monumental architecture or lifestyle demonstrations associated with consumption and social representation. Cities become ceremonial centers and transcended their original primary functions.

Indeed, the original movement of natural goods from the hinterland to the urban centers was reversed; the new consumption ideology and other urban "products" were transmitted to the hinterlands, on which the city used to depend. In the new paradigm, urban systems were recipients of goods, people, and ideas from areas beyond their hinterland, and this created a movement from the peripheries to the urban centers in the search of new opportunities associated with a "glamorous" lifestyle. However, in addition to more opportunities and the availability of more goods and services, urban life often meant the intensification of inequalities.

Privilege and power were further defined by the accumulation of wealth, made possible by the centralizing momentum of urban life, increasingly associated with economic and financial mechanisms dissociated from nature, physics, or basic human principles and values. This "artificial" urban life may be one of the aspects that has contributed to environmental, financial, and social crisis all over the planet, as globalization promotes the rapid spread of any disturbance in the system that characterizes modern economies.

Urban areas are, more than ever, the heart and soul of modern society, and therefore it is in the urban areas that we need to find the hope and energy for adopting a new paradigm that may promote sustainable development. Urban areas are therefore the cornerstone of sustainability, and whatever strategy we adopt for sustainability, it must grow from the original lessons of a fruitful functional relationship between the urban center and its surrounding area. In this context, we may consider urban areas as the lighthouses of sustainability (or unsustainability, if the trend is not modified).

In fact, this movement finds its roots in the seventeenth-century philosophy of Cartesian dualism—the dichotomous separation of humans from nature. The Cartesian dichotomy has reinforced ancient Western cultural expressions that placed humans above nature, as if we were neither interconnected nor interdependent. A major consequence of technological progress was that Western society was believed to be able to dominate nature and has become more and more disconnected from it, without realizing all the consequences of interacting with such a complex system. Human-nature dualism has proven to be one of the most important modern causes of human degradation of the biosphere, and it has produced many cascading effects.

In this context, the process of successful integration of urban and countryside should be based on both the memory associated with the cultural and historical identity of different urban areas, and on the diversity and dynamics that urban areas provide, and these contribute to our hope to develop and consolidate the new paradigms required to foster sustainable development.

In this, history shows that the metaphor provided by nature, and particularly by ecosystems and their functional rules, on which industrial ecology principles are based, may provide citizens with a better connection to natural resources and a respect for limits to growth and the idea of sufficiency.

If we view urban metabolism as based on the energy and material flows that are characteristic of the urban nucleus and the surrounding industrial and agricultural areas, as illustrated in figure 5.4, we observe that these flows do basically represent the consumption of raw materials and goods that are transformed by the industrial or agricultural areas or directly consumed in the urban nucleus. The products and goods produced by industrial and agricultural areas, or even in the urban nucleus,

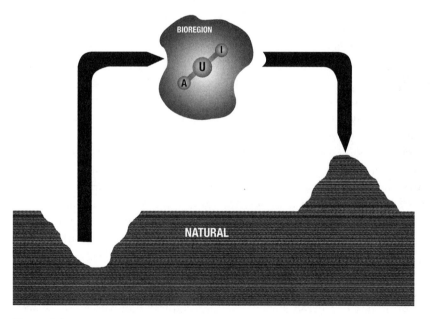

Figure 5.4
The urban metabolism as an open system.

can be consumed locally or exported to another urban system, however, most of the material flows are transformed in waste. Overall, the current urban systems are characterized by a massive consumption of nutrients (in an ecosystem language), such as food, water, fuels, construction materials, vehicles, or computers, and converted in waste, sewage water, or air pollution, thus configuring an open system, where materials flow in and out without major perturbations.

The first lesson that we may derive from industrial ecology principles is that the linear metabolism associated with urban areas, as represented in figure 5.4, needs to be transformed in a circular metabolism wherein wastes are converted back into nutrients (raw materials) that can be reintroduced in the manufacturing process by detritivores, which in our world may be represented by recyclers. This transformation is represented in figure 5.5, which illustrates the role of detritivores in transforming wastes into nutrients, which may feed the production processes, as in natural ecosystems.

In making use of the industrial ecology metaphor, it is important to realize that ecosystems are dynamic, composite entities where large quantities of matter, energy, and information flow within and between components in a way that is not yet completely understood. These flows depend on the ecosystem structure and determine it. There is not yet agreement as to whether the flows are controlled primarily by (Cury, Shannon, and Shin 2003):

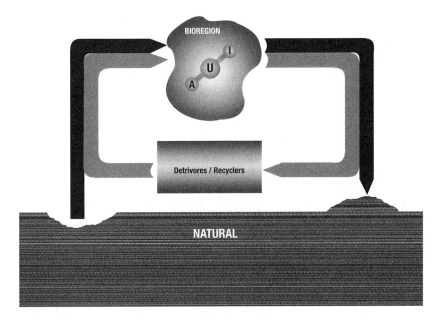

Figure 5.5
Closing the urban metabolism.

1. top predators' feeding behavior (top-down control);

2. primary producers (bottom-up control);

3. some numerically abundant species somewhere in the middle of the food chain (wasp-waist control); or

4. some combination of some or all of these, depending on systems and their possible states.

The lesson we can thus bring from ecosystems is that in order for urban systems to emerge as the cornerstones of sustainability, citizens and institutions should adopt a twofold strategy:

1. promote more sustainable consumption by reducing the amount and the hazardousness of material flows consumed, and

2. promote symbiotic relationships and recycling in order to contribute to closing the material cycles.

Making use of the industrial ecology metaphor, a successful implementation of both these strategies may require, as in natural ecosystems:

1. a major information infrastructure, based in adequate metrics and accounting tools,

2. new policies that build up adequate infrastructures to enable the closure of material cycles and sustainable consumption, i.e., a sustainability oriented governance with top-down control;

3. educational programs that may induce auto-regulated behaviors that may control the system within a more sustainable strategy, in a bottom-up control;

4. civil society movements, induced by leaders in sustainability that may bring together a significant number of citizens and induce public-private partnerships to enable a middle-out control model.

As in ecosystems, the solution will rely on a combination of these strategies, and therefore the key for success resides in developing the right vision, objectives, metrics, information, and above all education, as the pillars for sustainable development. In this process, success will also be measured by the capacity to develop within citizens a stronger sense of interaction, communication, and connectedness, as far as the industrial ecology might be embedded in the socioeconomic fabric.

Again, the urban systems have the adequate characteristics to start this movement, taking advantage of agglomeration, educational and communications infrastructures, scale, cultural heritage and memory, diversity or creativity.

Many national and local governments, supported by researchers, city planners, politicians, and the public are recognizing this great opportunity based on the growing understanding that urban areas are integral components of the planet's natural and physical systems, and that urban systems are whole systems that equal more than the sum of their parts. As a consequence, innovations for promoting sustainable development can be made by connecting the economy, environment, culture, and technology with sustainable practices and holding the future in mind.

In this strategy, resilient urban systems are recognized as important for achieving long-term urban sustainability (Resilience Alliance 2007). As discussed by the ministry for the environment of New Zealand (2008), urban resilience has two main characteristics: the first is the robustness or strength of an urban system to withstand stress. The second is the adaptability of an urban system to respond to changing conditions and objectives. Stresses can take many forms including sudden changes in environmental conditions, such as a major disaster, or economic shocks, such as the withdrawal of a major employer in the region.

One recognized way of creating resilience in an urban area is by designing a resilient infrastructure. Industrial ecology principles resulted from a very resilient framework, the ecosystems, and should therefore be relevant to the design of a resilient infrastructure, namely in promoting a greater use of more localized or dispersed and diverse ways of providing services like energy, drinking water provision, or food. An example of a localized service is the use of solar technologies like photovoltaic panels to enable homeowners to generate their own electricity. This sort of infrastructure can reduce the scale of damage from extreme events to more localized areas. For example, if an electricity supply cut occurs, resilient infrastructure design can restrict the outage to a few suburbs rather than the whole urban system, particularly if this is combined with smart energy network strategies.

Industrial ecology can thus provide major principles that might prove to formulate a strategy to promote a sustainable urban metabolism. A sample list of such industrial ecology–derived principles to design sustainable urban metabolism systems is presented in table 5.1.

Strategies toward a Sustainable Urban Metabolism

The establishment of a strategy toward a sustainable urban metabolism requires the definition of a roadmap, including a set of sequential and complementary phases that might be grouped in five essential steps:

1. establish a vision.

2. characterize the current urban metabolism.

3. establish an information framework that may support a sustainable interaction between humans, urban systems, and the environment, as well as the feedbacks from these interactions. A DPSIR framework is particularly recommended, in which the responses should provide clear policy goals.

Table 5.1
Major principles to formulate a strategy to promote a sustainable urban metabolism, based on the industrial ecology metaphor

Industrial ecology principles for urban systems	Comments
Systems oriented	Urban systems are a complex network of interdependent subsystems, as are ecosystems, for which the degree and nature of relationships is imperfectly known. The spatial and time scales of various subsystems are very different, and understanding of each individual subsystems does not imply global understanding of the full system. A holistic view focused on understanding system structure and behavior is required and involves building and managing transdisciplinary tools and metrics.
Urban metabolism	Urban metabolism provides a vision of the main material and energy flows that support and maintain the urban fabric by drawing analogy with the metabolic processes of organisms. This approach provides adequate tools and metrics to quantify the degree of circularity of resource streams, and may be helpful in identifying critical processes for the sustainability of the system and opportunities for improvement.
Respect the limits of nature	Economic growth must find ways to respect the limits of the natural system to provide the resources required for development indefinitely.
Sustainable consumption	Sustainable consumption requires a demand-side management approach that focuses on meeting genuine needs and on providing them with an adequate and efficient framework, and this is contrary to prevailing supply-driven systems that often limit or preclude sustainable options.
Close material cycles	Learning how to make better use of a waste stream, namely as a raw material for any other process, provides us with the possibility of substituting the term waste for materials that can be recycled, and thus transforming urban areas into mines for different types of material requirements.
Distributed	Natural systems are based on the provision of services and goods by a large network of a variety of distributed autonomous individuals, spatially dispersed, all of them acting, interacting, and evolving according to a limited set of spatial "laws of nature." In this context, the nature-inspired spatial metaphors provide a motivation for the promotion of distributed systems in opposition to centralized systems. The microgeneration in the energy sector provides such an example of the application of this principle at an urban level.

Table 5.1
(Continued)

Industrial ecology principles for urban systems	Comments
Resilient	Ecosystem resilience describes the capacity of an ecosystem to deal with strong disturbances and survive. A resilient ecosystem has the capacity to withstand shocks and, if damaged, to rebuild itself. Biodiversity plays a crucial role in ecosystem resilience by spreading risks through variety, cooperation, and redundancy that facilitate reorganization after disturbance. As discussed by the Resilience Alliance (2007), the resilience of urban systems recognize the role of metabolic flows in sustaining urban functions, human well-being, and quality of life; governance networks and the ability of society to learn, adapt, and reorganize to meet urban challenges; and the social dynamics of people as citizens, members of communities, users of services, consumers of products, etc,, and their relationship with the built environment which defines the physical patterns of urban form and their spatial relations and interconnections

4. develop adequate infrastructures.

5. establish a governance system that may facilitate a self-regulating, highly participatory system.

Vision

The establishment of a long-term vision based on unique distinctive features of a specific urban system is a critical element for the mobilization of the multiple actors that are involved in promoting a sustainable urban metabolism.

Establishing a successful vision is an inclusive and participatory process, bringing together different groups from the community such as public authorities, business, industry, academics, education, planners, civic leaders, environmental NGOs, and community associations, allowing for a vision that captures the different values, cultural communities, and interests of a broad constituency.

The U.S. Environmental Protection Agency suggests that this movement may include brainstorming ideas from the entire community, in a process that may result in a synergistic effect and bring out a myriad of ideas that reflect the values and interests of the community as a whole.

In any case, it is important that the vision highlight the specific characteristics of that particular urban system, considering the people's values, their needs, and the role of knowledge, culture, technology and innovation as ingredients to envision a sustainable development pathway that may benefit from the people´s ownership of the key message, as this will be a basis for genuine sustainability.

As an example, in CCDRLVT (2007, 83), Lisbon, the capital city of Portugal, has defined the following vision for 2020:

The region of Lisbon will become a cosmopolitan metropolis, relevant within the European capitals, fully embedded in the knowledge society and in the global economy, very attractive for its geographical singularities, qualities, Nature and Euro-Atlantic location. Social and environmental sustainability, the reinforcement of socio-territorial cohesion, the promotion of ethnical and cultural diversity and efficient governance are, in that time horizon, prerequisites and goals of the economic and social development goals in the Lisbon region.

This vision is focused on the people and based on four major strategic axes:

1. *Competitive region*   Lisbon will emphasize the quality of its territory and people in promoting international intermediation activities with a high technology profile.

2. *Cosmopolitan region*   Lisbon will take advantage of its history and culture as a locus of privileged Euro-Atlantic interaction to continue promoting itself as a place of excellence to host multiple cultures and civilizations and promote the value of solidarity.

3. *Cohesion*   Lisbon will invest in people, with relevance to promoting education, high professional qualification, and scientific, cultural, and social jobs.

4. *Connectedness*   Lisbon will assume a major investment into the modernization of administration and institutional solidarity, into the innovation of governance processes and citizen participation, into the control and monitoring of results of investments made, and above all, in those that will contribute to promote added value based on a systems view of a networked urban area.

The establishment of the vision provides a direction, but in order to design a more precise roadmap we need to characterize the point from which the system is departing, and an adequate strategy to reach the goals defined.

Characterize the Current Urban Metabolism

The characterization of the current situation of an urban system, and ultimately of the urban metabolism, requires the identification of a battery of indicators that normally may be classified in the drivers, pressures and state categories of the DPSIR framework analyzed in chapter 1. Examples of these indicators are provided in table 5.2.

In any case, the selection of an indicator to the framework of selected indicators for the urban system management needs to follow a set of predefined criteria. Here we adopt a methodology based on that defined by the Organization for Economic

Table 5.2
Sample drivers, pressures, and state indicators in urban systems

Indicator type	Theme	Indicator
Drivers	Population and territory	Number of inhabitants
	Energy	Electricity (in KWh) and gas consumption (m^3) per capita
	Transportation	Park fleet (vehicles per 1,000 inhabitants and vehicles per km^2)
	Education	Birth rate (babies/year)
Pressures	Global warming	Average CO_2 emissions per inhabitant per year
	Transportation	Number of vehicles per hour in a given location
	Education	Number of students enrolling in primary education
	Waste management	Production of waste per capita (kg/day.capita)
State	Global warming	Atmospheric concentration of greenhouse gases in the air (parts per million, ppm)
	Urban air quality	Atmospheric concentration of SO_2 and NO_2 and in the air (parts per million)
	Education	Percentage of students reaching the ninth grade
	Waste management	Percentage of waste recycled

Cooperation and Development approach, which uses three selection criteria: political relevance, scientific soundness, and measurability.

The political relevance of an indicator refers to its usefulness for securing the fulfillment of the vision previously defined. As discussed by Alberti (1996), the indicators should also provide a basis for international comparisons and have a target or threshold against which to compare environmental quality and performance.

The scientific soundness of indicators requires them to be based on theoretically well-founded studies in technical and scientific terms. For example, the urban metabolism indicators should be based on material flow analysis studies, which, as discussed in chapter 4, are based on international standards and have international consensus about their validity. The measurability of indicators requires data to be readily available at a reasonable cost-benefit ratio, adequately documented and of known quality; and updated at regular intervals.

A DPSIR Framework
The DPSIR framework provides information to support a sustainable interaction between citizens, urban systems, and the environment, as well as the feedbacks from

Table 5.3
Sample impacts and response indicators in urban systems

Indicator type	Theme	Indicator
Impacts	Population and territory	Unemployment
	Energy	Global warming (kg CO_2 equivalent)
	Transportation	Congestion (time spent in traffic)
	Education	Rate of abandonment of school (%)
Responses	Population and territory	Unemployment compensation (€/month)
	Global warming	Energy efficiency in buildings (kwh/m^2)
	Transportation	Investment in public transportation (€/capita)
	Education	Number of schools per inhabitant

these interactions. The D-P-S part of the framework has been analyzed above, and the full framework requires the recognition and the quantification of the impacts that are observed as a result of the pressures caused by human activity and subsequent policy responses, defined in the form of quantified goals, making use of adequate indicators. This constitutes the I-R component of the framework, which provides the political feedback. Examples of I-R indicators are presented in table 5.3.

Such a framework of indicators constitutes a unique tool for policy making, as it provides a model that facilitates the understanding and the public perception of complex interactions between drivers, their actions, and impacts and the responses that may improve the global system's sustainability.

In fact, an urban DPSIR framework serves different purposes: (1) systematic and coherent identification of the different dimensions of urban sustainability; (2) monitoring of the evolution and interaction of the major indicators; (3) target setting; (4) evaluation of the efficiency of the policies adopted to mitigate the problems detected; and (5) public information and communication.

Although an indicators framework constitutes a simplified model of an urban system, it provides a vast amount of information that is easy to read and understand. In fact, it permits, for example, correlating the level of emissions of a particular atmospheric pollutant to complex environmental phenomena, such as acidification or climate change, and allows policymakers and the public to assess policy performance over time.

Develop Adequate Infrastructures

The importance of the strategic plan cannot be overemphasized, in which both short- and long-term city performance goals and objectives need to be identified, as well as the trajectory supported by an adequate infrastructure. In the light of the industrial ecology metaphor, some key requisites to the infrastructure design are

- resilience, in the sense that the city can respond to any stress and, if damaged, rebuild itself;
- flexibility, in the sense that the city can respond to nonanticipated challenges; for example, to install in a building an infrastructure for hot water distribution that will enable future use of solar collectors;
- distribution, enabling, for example, distributed energy sources, normally associated with energy microgeneration, e.g., at the building level;
- smart, in the sense that we should enable the infrastructure with the maximum information availability, as for example, through the use of smart meters that may provide instantaneous information on the energy consumption in energy-consuming appliances;
- communicative, in the sense that we should provide ease of data transfer between any equipment that might use the infrastructure. For example if we have smart meters in electrical appliances, this will be of no use if they cannot transmit that information;
- able to facilitate the use of renewable natural resources;
- able to facilitate human interaction and learning experiences;
- able to foster connectedness, to facilitate the establishment of networks;
- able to facilitate respect for the limits of nature.

Establish a Governance System to Facilitate a Self-Regulating, Highly Participatory System

Governance will ultimately determine the rate of progress toward sustainable development and will provide the institutional guidance for the goals established through the response indicators in the DPSIR framework.

The industrial ecology metaphor suggests the observation that ecosystems have no central controlling body or mechanism, but their various parts work together to make them support life at local, bioregional, and global levels, as discussed by Newman and Jennings (2008). In the natural system, regulation springs up internally through feedback loops facilitated by proximity and network relationships. The "governance" system is decentralized but strategic for the different levels at which its life forces are dependent and to which they contribute, and this allows for flexibility in responses.

Newman and Jennings (2008) suggest that there are five main ways for cities to create good governance and hope. All need to be given the particular flavor of hope that seeks out truly positive, achievable steps that can be taken to launch cities down the path toward sustainability:

1. structures and processes of urban governance need to be inclusive, cooperative, and empowering in a way that reduces inequalities;

2. governance needs to be matched to local and bioregional scales, but have the ability to address global issues. Governance structure and processes need to facilitate visioning processes and to foster empowerment and partnerships;

3. sustainability needs to be embedded in the day-to-day operation of government;

4. indicator projects and reporting need to be developed;

5. governance structures need to support and facilitate the flourishing of community initiatives for sustainability, providing a wellspring of hope for the future.

The faster we adopt these guiding rules based on urban metabolism concepts, the faster we will change our paradigm to one of sustainable development. Having the tools, the adequate frameworks and indicators, available, it is in our hands to promote change, and this is emerging as discussed in the next chapter.

III

Sustainable Urban Systems

6

Green Urban Policies and Development

The development and implementation of policies aimed at promoting sustainable cities has accelerated in direct response to the intensification of the need for practical strategies to increase the resource efficiency of contemporary cities. Green city initiatives, urban environmental action plans, solar and renewable city goals, zero energy and carbon emissions developments—all are indicative of a frenetic interest in promoting the transformation and growth of sustainable urban zones. Despite the intense interest and significant financial and intellectual resources directed toward these efforts, it is clear that these policies have been created and implemented within a context of incomplete understanding of the fundamental behavior of urban production and consumption over time. Even those cities that have embarked on multifaceted, data-driven green initiatives are working toward their goals without much precedent or knowledge of the likelihood of success. As a result, green city initiatives are becoming one of the most pervasive and extensive resource-management and social experiments ever conducted without the benefit of clear and extensive data. The possibility that significant savings in resources will actually be achieved while promoting humane and sustainable contemporary cities is a matter of debate. To date, there is little evidence to suggest that actual reduction in consumption, increases in resource efficiency, or improvement in our ecological footprint and carbon intensity are occurring. Chapter 6 will review green city initiatives.

Emergence of the Green City

Cities have always maintained an intimate and sometimes uneasy relationship with the countryside. Since the earliest settlements, the world immediately beyond the perimeter of city walls provided much of the physical needs of the urban population, yet also harbored the dangers of competing tribes, the unknown of the wilderness, and—most haunting of all—a lack of societal structures and community. However, the settlement could not survive without the provisions supplied by agricultural land

and domesticated animals, as well as sources for metals, minerals, water, and other critical resources found beyond the secure urban perimeter. The "hinterland" provided these goods and depended on a central urban administration and resident militia for security and governance. While this mutually beneficial relationship has always been critical for the economies of both the city and the countryside, the long-term effects of urban consumption have been asymmetrically favorable toward the needs of the agglomeration economy.

As cities have grown and become enormously wealthy engines of concentrated innovation economies, the hinterland has been diminished as resources are extracted and wastes are dispersed. At least in terms of natural resources, the city has always depended on the countryside much more than the countryside needs the city.

Today this uneasy symbiosis extends to the global "hinterland"; that is, the un-urbanized portion of the world that provides much of the biomass, minerals, fuels, water, clean air, and other critical resources that all cities and their residents depend on through regional, intranational and international trade. The source of unease with the countryside stems from the inevitable need for outright exploitation of the resources of the immediate and distant hinterland that has characterized the settlement and growth of every city everywhere. There has been no urban condition in which the adjacent hinterland has not been affected to some extent, and sometimes with devastating environmental consequences. The extension of this unease to a global scale is partly due to the powerful intensity with which cities draw upon the critical resources that only the global economy can provide, by way of international flows of finance, freight, information, and people. This global urban-hinterland relationship is similarly asymmetrical, feeding local urban economies while drawing down global natural capital and degrading environments that may lie many thousands of miles away from the source of urban consumption.

This is not a new phenomenon. In fact, the degradation of the countryside due to the founding and growth of an adjacent settlement is a seemingly inevitable, or at least exceedingly common, result. Examples of environmental damage and problems arising from degrading and exhausting local resources can be found in ancient settlements and cities from around the world. Research has found that as far back as the beginnings of Neolithic settlements, small villages of only several dozen inhabitants had an important effect on the environment of the surrounding countryside. These effects can still be found in the countryside in which these settlements were located (Comer 2003). Ancient Greece suffered through vast deforestation and soil degradation as a result of agricultural cultivation and grazing of domesticated animals as far back as 4,000 B.C. (Runnels 1995). Studies of ancient Ur suggest that the city was responsible for the deforestation of a large area of Mesopotamia and experienced shortages of water due to intense agricultural cultivation. Similarly, the forests of China have been exploited for construction and fuel for many hundreds of years to such an extent that, as of 200 years ago, no original forests remain in the entire country (Ponting 1991).

The degradation of the ecology and environment of the hinterland has not been the only consequence of urbanization through the ages. In fact, urban growth has not only affected the environmental conditions of the hinterland—it has equally compromised the urban environment and ecology. This is mainly due to the fact that, through the ages, cities have been consistently effective at acquiring resources that can be mobilized from near and far and directing these resources to converge on a city center, while demonstrating significant deficiencies in dispersing wastes away from the urban space.

As the Industrial Revolution transformed national economies, cities were unable to adapt quickly enough to mitigate an explosion of pollution and waste accumulation within urban boundaries. The environmental conditions of nineteenth-century London are well known. A combination of urban industrial water and air-borne effluents and the soot generated from the use of coal to heat homes resulted in a caustic and unhealthy urban environment. The accounts by many writers of the time document the conditions of the inner city. Charles Dickens provides one example:

It was a foggy day in London, and the fog was heavy and dark. Animate London, with smarting eyes and irritated lungs, was blinking, wheezing, and choking; inanimate London was a sooty spectre, divided in purpose between being visible and invisible, and so being wholly neither. (*Our Mutual Friend*, 1877)

Until very recently, typical urban environmental conditions in the industrialized, developed world were quite awful, with terrible air and water quality, waste management problems, and other pollution issues leading to serious health issues for urban residents. During the nineteenth and early twentieth centuries, the acceleration of the global economy fed by the unprecedented extraction of natural resources, particularly fossil energy carriers, created urban pollution problems that seemed intractable. Early twentieth-century Boston, New York City, Pittsburgh, and many other industrialized cities were contending with woefully inadequate solid waste management systems coupled with unfettered license for industry to exhaust caustic gases and particulate matter into urban air and water.

Clearly, air and water resources have suffered most by the willful discharge of urban wastes, and have lately shown evidence of greatest improvement from the management and reduction of these effluents. Addressing these flows effectively can certainly lead to positive short-term results. This has been one of the lessons of urban environmental management and the associated regional and national policies aimed at addressing the environment.

In the United States and elsewhere, national, regional, and urban policy development and implementation of air and water quality statutes has yielded dramatic and surprisingly rapid improvements to seemingly insurmountable urban pollution problems. Los Angeles is a good example of the quick improvements that can result from policies aimed at reducing urban pollution. The passage of a series of laws known collectively as the Clean Air Acts of 1963, 1970, and 1990 and stringent local pollution regulation within the L.A. basin both acted to set into motion dramatic reductions in the pollutants, including particulates, arising from industry and automobile exhaust. In short order, the air quality of Los Angeles improved to levels not seen since the mass adoption of the automobile for personal transportation. Between 1980 and 1998, the levels of sulfur dioxide and ozone both fell by 9 and 12 percent, respectively. Carbon dioxide, nitrogen dioxide, and particulate matter also fell. Overall, the air quality of the Los Angeles basin has improved and been sustained as a direct result of these regulatory statutes. Importantly, the reductions in urban air pollutants have occurred during a period of increases in urban population and economic activity coupled with an increase in per capita automobile use and a decrease in per capita mass transit use.

However, some problems resulting from a misunderstanding of the dynamic of natural urban systems are not so easily rectified.

Possibly most problematic and difficult to analyze is the effect of diverse land uses and actual urban physical form on the resource intensity of the urban settlement. It is clear that the urban heat island effect, the lack of and diminishing quality of urban water, the degradation of biodiversity, and other insults on the biogeochemistry and ecology of the urban environment are a direct result of the actual physical form of a city, such as size and use of buildings and impermeable surface. What is not yet clear is the quantitative link between distinct urban patterns—morphologies—and the resource flows associated with it (Alberti 2005). More work is required to make this link. In particular, Alberti has concluded that researchers need to better establish the relationship between specific urban attributes of compactness, density, connectivity, and heterogeneity (Alberti 2007). Some preliminary evidence indicates there are strong correlations between these attributes and ecological function, but these correlations are highly dependent on the location of the city and its climate, population, and population density, among other variables.

Several researchers have also attempted to make links between urban growth patterns and the accompanying effects on the biogeochemistry and ecology of the urban region. These links have been explored as a result of land conversion from natural and agricultural uses into urban land and from low-density sub-urban settlements into high-density central city developments. Strong correlations have been made between population and housing densities, with transportation intensity as a resultant of the two (Pataki et al. 2006).

For example, several studies have pointed to direct correlations between the number and diversity of bird and mammal species with respect to an urbanization gradient from city center to pericenter and suburban to rural. Results seem to indicate strong negative correlations between biodiversity of key species and urbanization, with some exceptions for those high-density cities that also incorporate substantial green areas within the city center (Clergeau et al. 1998; Clergeau, Jokimäki, and Savard 2001; Snep 2006; Hepinstall, Alberti, and Marzluff 2008).

There remains work to be done in arriving at strong quantitative links between urban form, transportation network configurations and technologies, and the resource intensity and carbon emissions of those settlements, though many projects are underway. In addition, there is important work left to do in identifying the essential variables that determine resource intensity and carbon emissions between like cities located in different parts of the world, in developed versus developing countries, or in contrasting climates and topographical conditions.

Clearly, there is substantial analytical ground to cover in establishing the connection between urban planning and design, infrastructure design and construction, or transportation network engineering and the resource consequences of urbanization generally. Effects of urban production and consumption that can be measured directly as a result of industrial processes and consumption of energy and materials is relatively straightforward compared with the correlations sought between the physical form of the city and resources.

However, the importance of this work cannot be fully appreciated until we know more. For example, retrofitting the existing city, its structures, and infrastructure toward much greater efficiency in consuming the resources that are expected of urban living and working is an important step toward urban sustainability. In addition, the ramifications for urban resource consumption as related to the physical form of the city are a fundamental aspect of understanding urban growth. This is especially important in those areas of the world in which accelerated massive urbanization is predicted. It is in evaluating the resource burden of future urban growth that more effective pathways toward societal sustainability will likely be found. Therefore, correlating resource flows to urban growth is an essential aspect of the study of urban metabolism.

Despite the strength of the successes of regional and national pollution control and urban waste management policies, the central catalyst for the contemporary drive toward urban sustainability has emerged from global concerns regarding continued access to critical resources and the uncertain consequences of global climate change. These concerns, while complementary to urban waste and pollution management issues, pose enormous additional challenges to be addressed by cities continually striving for economic development and contending with growing populations. Cities that intend to act today cannot wait for definitive explanations and scientific models that precisely articulate the relationship between urbanization and resource extraction, or processing and consumption. A number of cities have proceeded in the best ways they can, with limited tools and incomplete data.

Making Sustainable Urban Places

Despite the serious lack of essential knowledge regarding urban sustainability, the general notion has been a useful catalyst for generating interest and initiating diverse projects toward greater urban environmental responsibility, resource efficiency, and resilience. Energy efficiency, material recycling, urban waste management, and so on are all central concerns of municipal stakeholders and residents exploring best practices for manageable growth. In addition, sustainable economic development that explicitly addresses urban social issues is an essential attribute of urban sustainability.

The notion of sustainability itself is intimately related in its origins to the broader challenge of sustainable development globally. As embodied in the twenty-seven principles of the Rio Declaration on Environment and Development, priorities of sustainable development mandate "a healthy and productive life in harmony with nature" while "eradicating poverty as an indispensable requirement for sustainable development" that leads to "economic growth and sustainable development in all countries" (United Nations 1992). Urban sustainability is the localized embodiment of the global scale of these principles and is thus concerned with both wide-ranging resource issues and city-specific ongoing challenges of social and economic conditions.

The intertwined nature of urban resource and social priorities is a daily reality for urban communities intent on moving toward sustainable practices. The debates preceding community actions for urban sustainability place in high relief the mix of local cultural, economic, and social factors that must be reconciled with the best strategies for achieving real gains in reducing the resource footprint of the community—whether it be a neighborhood, a city center, or a large municipal region. This local reality is also the source of strength for the best prospects for urban sustainability. Some municipal sustainability plans are clear examples of *thinking globally* while *acting locally*. Mainly due to the fact that urban zones are limited in size (compared with nations) and are governed by various forms of central municipal authority, it is possible to affect some kinds of change in an effective way. As the Brundtland report states:

Local authorities usually have the political power and credibility to take initiatives and to assess and deploy resources in innovative ways reflecting unique local conditions. (World Commission on Environment and Development 1987).

These local authorities may embark on a program for addressing environmental and resource issues within their boundaries in a variety of ways. Many cities have cobbled together a mixture of private-public partnerships sometimes, including nongovernmental organizations, in exploring the possibilities for greater urban sustainability as a precursor to developing a city sustainability plan. The motivation to begin this process varies with location and often includes local political realities and tendencies, land use and critical resource issues (such as water shortages), transportation and built environment inefficiencies, and often a complex combination of other social, cultural, and economic trends.

Several organized movements toward codifying priorities for environmentally progressive urban zones have started in cities in which *nature* is at close proximity. Portland, Oregon, is an example. Situated at the edge of a vast natural world, Portland's residents live and consume in a large modern and progressive metropolitan area situated within a region of astounding national beauty and biodiversity. Another example is Cape Town, South Africa, located at the southernmost tip of the African continent between mountain and sea; city residents are reminded daily of their place in the natural world.

The perception of the human-nature interface creates and sustains the popular and political momentum to address the relationship between the city and nature. Urban sustainability movements have been catalyzed by the human-nature interface, most notably in cities where this interface is a daily reality: Seattle, Washington; Austin, Texas; Santa Monica, California; and Boulder, Colorado, in the United States—cities surrounded by a wealth of natural biodiversity and beauty.

Various drives toward urban sustainability have inevitably come to adopt diverse priorities and methods reflected in policy, engineering and design at a variety of urban and regional scales. Many examples of this diversity as built cities and ongoing development projects now exist: Huangbaiyu and Dongtan eco-cities in China; Masdar City in the United Arab Emirates; Waitakere, New Zealand; Gothenburg and Älvstaden, Sweden; the UK Eco cities initiative, and others. The diversity of approach is understandable. Differences in fundamental factors such as climate, topographical and other geographic variables, economic attributes and level of urban affluence, environmental issues arising from the history of industry within the region, and cultural and political frameworks require contrasting approaches and definitions of urban sustainability.

This diversity is still being augmented by the enormous growth in interest developing around the idea of sustainable cities. However, there are now several good examples, useful case studies that give an important indication local urban efforts toward sustainability can be effective.

Thousands of cities around the world have decided that the time to act is now. Of the many attempting to move toward urban sustainability, several of those efforts are described below.

The Nature of Green City Policies

Green city policies are driven by priorities that target specific local environmental and resource issues while striving to measure up to global priorities and principles.

A diverse mix of goals generally characterizes the content, development, and implementation of green city plans in many parts of the world. These plans are highly modulated by local conditions: neighborhood identification and extent, infrastructure engineering and conditions, topography and other natural and geographic features, etc. These plans are often highly motivated by visions of a future in which symbiotic relationships between human activities and natural cycles can be fostered and celebrated for the benefit of current and future urban residents. These visions are often severely tempered by many of the same longstanding issues that characterize the complexity of problem solving in the urban space. Three issues in particular are worth noting here: financial and organizational limitations, the physical and temporal aspects of "lock-in," and the challenges of urban poverty, health, and education.

The first issue is a fairly common limitation shared by almost every city in the world: a serious lack of financial resources and organizational efficacy. Contemporary cities in every country of the world are administered under tight and sometimes critically constricting financial limits. In many nations and states, cities are limited in their power to levy, assess, and collect taxes. Much of the overall revenue generated from taxes flows straight to the state or province and the central national seat of government. While urbanization and development often lead to the potential for broader municipal tax resources, the costs of providing for urban services often surpass these new revenue streams. Competing for firms and workers with other cities accelerates the race to provide incentives for these new companies, while urban competitiveness has to be considered a part of the global marketplace. Firms now relocate and expand into regions and to cities that can provide the services and amenities most attractive to running their businesses and attracting workers. Providing the services necessary to attract and maintain a thriving economic base has become more challenging to many cities around the world.

The financial limitations of municipal governments also affect their ability to employ experts in agencies that can effectively address the intricately woven issues of environment and economy. Even the largest cities in the United States with aggressive sustainability plans are often challenged by a lack of resources to staff effective agencies with real power to effect change. Many urban sustainability initiatives in the United States are expansions of the scope of work carried by municipal housing, transportation, energy, water, sanitation agencies, and so on. This additional work burden not only limits the extent to which government workers can become deeply engaged in an effective transformation toward urban sustainability, but also assumes that the expertise required can be easily and quickly acquired by already overburdened city workers.

Second, cities are large, physical assemblies of buildings, highways and roads, infrastructure networks, natural ecologies, and other extensive and generally immovable things. While it is tempting to imagine all manner of the "greening" of the urban landscape accompanied by the deployment of urban wind turbines, photovoltaic arrays, vertical urban farming, and low-energy personal urban vehicles, the reality of urban transformation must contend with the entropy and inertia that settles in because of the large scale and physically intensive commitment required of many of the most promising re-planning ideas. At many spatial and temporal scales, there is a certain "lock-in" factor that tempers most urban transformations, even those with the lightest touch.

For example, at one end of the scalar spectrum, the location of a city is, essentially by definition, unchangeable. Cities are not moved from one place to another. If a population is displaced as a result of natural disaster, dam building, or economic atrophy, the city itself is not considered to have moved from its original location to another. The untraveled roads remain, abandoned buildings deteriorate and eventually collapse, and the natural ecology begins to subsume all, but the city remains. As a result, aspects of the city that are a result of its location, climate, geography, geology, and topography, among other factors, are locked in, unchangeable. Sustainability plans cannot provide visions that counter attributes of the city that are locked in.

At the other scale of this issue are elements of the urban context that are unchangeable, but only at enormous cost and consequence. For example, since the 1950s and until recently, an elevated central highway dividing the financial district from the North End carried north–south traffic through the heart of Boston. The elevated highway had been envisioned as early as the 1920s and was engineered and constructed as part of post-WWII transportation planning in Massachusetts. Requiring significant neighborhood disruption and reconfiguration, the highway was part of the downtown Boston urbanscape until the city agreed to construct a series of tunnels directly below and remove the elevated highway. The Central Artery/Tunnel Project, known as the "Big Dig," was completed in December 2007 at a cost of over $14.7 billion, though it is likely the total may eventually exceed $22 billion due to interest payments on project financing. The half-century life span of this enormous piece of urban infrastructure is an example of the very real possibility to radically transform even the largest elements of the city, but at a very high cost. Many urban transformation projects of much smaller scale are not realized due to cost and other challenges and therefore represent a kind of lock-in of the urban form.

Lock-in is, admittedly, only a general concept of limited application value in assessing urban sustainability potential. However, there is a substantial literature on the relationship of cost to flexibility, especially in a context of uncertainty, in the design and engineering of complex systems; see figure 6.1 (Dixit and Pindyck 1994; Amran and Kulatilaka 1999; Ford, Lander, and Voyer 2002; Leviakangas and Lahesmaa 2002; Flyvbjerg, Bruzelius, and Rothengatter 2003; Ho and Liu 2003;

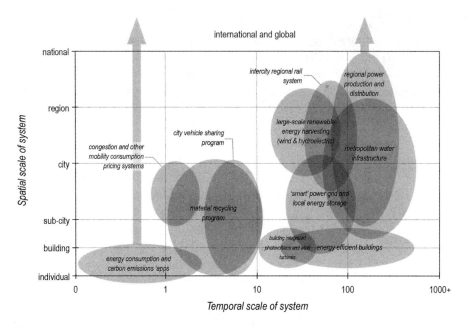

Figure 6.1
Spatial versus temporal scale of lock-in for a range of urban transformations.

Greden 2005; de Neufville, Scholtes, and Wang 2006; de Neufville and Scholtes 2011).

The third issue is the ongoing challenge of providing services to urban populations. In light of the dynamics of population emigration and migration, aging, household composition shifts, and other social and cultural trends, cities are more often challenged to meet the reliable delivery of basic services, such as employment opportunities, poverty alleviation, adequate health services, and education. Today, many cities around the world cannot deliver even minimal services for sustaining the lives of vast numbers of their residents. The urban poor in developing countries suffer the most, but many urban residents in developed regions also want for lack of clean water, reliable and affordable energy, and access to basic transportation services.

Thus, while the aspirations toward sustainability are found in the halls of many municipal governments and espoused by hundreds, if not now thousands of mayors around the world, the resources to achieve even a passing likeness of sustainability under the weight of current challenges are extremely limited.

Therefore, the derivation of urban sustainability measures and the formulation of green city plans must be considered within this context. The following examination of green city plans assumes this state of contemporary urban affairs.

As described in a study by Kent Portney, sustainable city initiatives commonly include a limited set of "activities" that are meant to contribute in various ways to the multifaceted goals of the overall urban sustainability initiative (Portney 2001). Portney lists the following as key activities:

1. sustainable indicators project: measures of sustainability that indicate the progress the city will highlight, eventually leading to the development of actual targets and benchmarks to be achieved within a specific period of time.

2. controlled growth plan: delineation of a preferred growth pattern and extent that is sensitive to important resource and ecological elements of the urban environment while promoting certain specific urban form and density goals, such as establishing an outer urban growth limit with disincentives for development outside of its boundary.

3. sustainable transportation planning: often intimately linked to the controlled growth plan, a city may focus explicit attention on the variety of transportation modes that are supported by the city, including investments to upgrade and expand mass transit access, pilot programs to provide shared transit resources such as city bikes and scooters, and financial incentives for public transportation and disincentives for traveling by automobile (see figure 6.2).

4. pollution control and waste management: mapping the pathways of municipal pollution and waste while organizing and assessing strategies for controlling its volume and composition.

5. energy and resource management: provision for the production and purchasing of all kinds of renewable energy for household and industrial consumption while focusing on the low-hanging fruit of conservation of water and energy, especially in the built environment.

6. global climate change: lately included in many sustainable city plans, the risks of global climate change are addressed by cities in many ways depending on the location, vulnerability, political will, and economic health of the city.

These six activities define the core elements of many of the green city plans developed by cities to date. The listing below highlights major urban sustainability efforts that are currently underway.

Green City Initiatives

Portland, Oregon

Portland, Oregon in the United States has been at the forefront of a nascent green city movement since its early investment in light rail. It has made efforts to increase urban density and prevent sprawl through land-use planning, attention to center-city

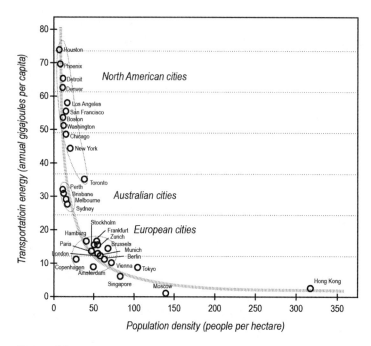

Figure 6.2
Transportation energy and urban population density. Another aspect of lock-in is represented here as population density serves as a proxy for the density of the road network. (Sources: Newman and Kenworthy 1988; Bournay 2008.)

urban design and streetscape renewal, and various creative measures to curb pollution and manage municipal wastes. Located in the northwest United States on the Willamette River, Portland is possibly best known for its experience in adopting a state-mandated limited growth policy. The Urban Growth Boundary (UGB) was a land-conservation and urban-density initiative championed by the state of Oregon and its governor Tim McCall in 1979. Portland has been a case study in the successes and pressures of promoting urban density while restricting development on surrounding agricultural and natural lands (see figure 6.3).

The results of the UGB and the associated effort to renew the downtown area beginning in the 1970s are a clear indication of the complementary and mutually reinforcing relationship between regional planning efforts to control low-density urban land expansion and local efforts to promote and support dense mass transit options and housing development.

Austin, Texas

Austin, Texas contrasts with Portland in that its resource-efficiency initiatives come substantially from within the boundaries of the city. While the state of Texas has

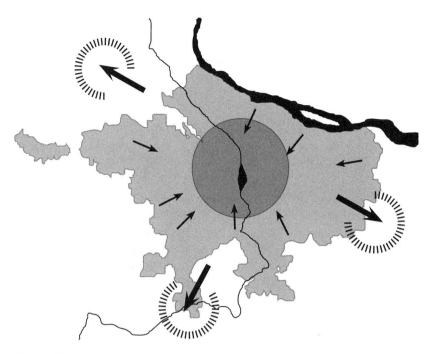

Figure 6.3
The urban growth boundary of Portland, Oregon. While the city has generally worked toward maintaining the integrity of a spatial bound for new development, future areas of expansion along transportation and topographical corridors outside of the no-growth boundary are being studied and approved. Simultaneously, the city is working toward greater density in the center of the urban space.

shown some commitment to renewable energy, notably wind energy, it cannot be cited as an environmentally progressive state. The conservative politics of the state do not necessarily align with the liberal culture of Austin, and therefore regional and state priorities do not often support the progressive municipal political climate. Yet Austin has shown a deep and sustained commitment to the environment and resources in diverse ways that today include mandatory energy audits as a requirement for every home sale; substantial support and subsidies for renewable energy; information, training, and resources for water conservation; recycling; biodiversity and habitat protection; and pollution reduction and waste management.

Austin is located in the central region of the state, with a 2011 population of approximately 767,000, an increase of 17 percent since the year 2000. Austin is home to the state government and the University of Texas. Of central importance to a sustainable vision for Austin and its role in protecting and residing responsibly within the central Texas region has been the explicit spatial identification of areas for growth and areas for protection from urbanization. Adopting many of the central tenets of smart growth, Austin has designated a zone for growth called the desired development zone (DDZ) and a zone of restricted development called the drinking water protection zone (DWPZ). These zones are both identified as part of the Smart Growth initiative. In addition, the city actively supports increased mass transit ridership through its Transit-Oriented Development initiative.

Toronto, Canada

Toronto, Canada, also provides evidence that much can be accomplished primarily through local means and effort. In particular, Toronto has made a strong effort to take specific actions to adapt and mitigate the effects of climate change at the municipal level.

Recently, Toronto has begun to see the results of a law passed in 2010 promoting the installation of green roofs. In the near term, additions amounting to approximately 36.5 hectares are planned for commercial and residential buildings in the city. The city was prompted to pass the law based on the generally accepted understanding that planted roofs assist in the management of run-off and reductions to the volume of storm water while also significantly contributing to reduction in the urban heat island effect. A simulation by Environment Canada of 93 hectares of planted roof in the climate and urban density of Toronto predicts that a reduction of 1 to 2 degrees Celsius and an associated energy savings of 15 million kilowatt-hours could be anticipated. As this book went to press, a green roof has been completed on an entrance pavilion of the Toronto City Hall (Urban World 2011).

Cape Town, South Africa

Cape Town has been a leader in Africa for advanced, comprehensive, and practical policies aimed toward a sustainable urban future. Two aspects of Cape Town's move toward urban sustainability are highlighted here: the sustainable procurement policy and the *Smart Living Handbook*.

Ensconced as part of Cape Town's Environmental Agenda 2009–2014, the Green Procurement Policy satisfies the intent of one of the environmental principles, stated as, "Entrenching sound environmental values and responsibility within all aspects of society, governance and decision making" (City of Cape Town 2003; 2009, 2; 2011b). While there are many aspects to the environment agenda, the procurement procedures are most closely aligned with an explicit focus on the urban material flows necessary for urban respiration; at least that part of respiration that is under the control and management of the city government and its vendors.

Cape Town has been aggressive in targeting the savings and environmental improvements that can be achieved through a resource-efficient supply chain management policy. The policy, approved in December 2011, stipulates detailed guidelines that are meant to, "ensure sound and sustainable accountable supply chain management within the City of Cape Town, whilst promoting black economic empowerment, which include general principles for achieving … socio-economic objectives." Of particular interest here is the explicit mention of objective 6.2, "to promote resource efficiency and reduce the negative environmental impact of daily operations of the City": and to link this kind of awareness of resource flows with socioeconomic objectives.

The second highlight of Cape Town's efforts is the intensive outreach program aimed at households and consumer lifestyle behavior. The Smart Living Handbook is a comprehensive and impressive collection of practical guidelines toward resource-efficient urban living (City of Cape Town 2011a). Organized into four sections, waste, energy, water, and biodiversity, the SLH outlines these resources at the scale of the city before it offers a perspective from the scale of the household. SLH provides these various perspectives to illustrate government's role in providing and managing these resources, the role of business and entrepreneurial initiatives, and the individual consumer's role in their own household. Simple checklists are provided for the consumer to track their consumption and manage their own household material flow.

Detroit and New Orleans: Sustainability after Deurbanization and Trauma

In addition to city sustainability plans prompted by the growing awareness of the need to more carefully manage urban consumption, some urban planning has been the result of dire environmental and economic circumstances. Two examples from the United States, that of Detroit, Michigan, and New Orleans, Louisiana, are particularly illuminating regarding the de-urbanization of large areas of the interior of the country, especially related to employment losses in heavy industry and agriculture, and dramatic population shifts due to environmental and natural disasters. These latest initiatives are different from others in that their origins and focus are toward rapidly deteriorating critical situations.

Detroit, the poster child of shrinking U.S. cities, is undergoing a dramatic transformation. During the first decade of the twenty-first century, Detroit, an emblem of the manufacturing might of the United States in the twentieth century, lost 25 percent of its population. Since 1970, the city's population has decreased by 57 percent. In contrast, the extent of the greater metropolitan footprint has continued to expand, leading to a dramatic decrease in the overall density of the city. That is, while Detroit is certainly a 'shrinking city' in terms of overall population, it has not reduced but rather extended its spatial extent.

The city is now in the process of working through the social, economic, and logistical issues of actively promoting population shifting from more blighted neighborhoods to less. Continuing to provide the same levels of service in substantially depopulated areas is not only financially expensive but resource wasteful. Of course, this is an easy observation to make and a difficult one to act on.

Other Green City Initiatives

Lately many more cities, too many to list here, have begun to adopt programs that explicitly address resource efficiency, urban environment, alternative infrastructure, and global climate change.

On the West Coast of the United States, the cities of California, San Francisco, Santa Barbara, Santa Monica, Los Angeles, San Diego, and others have developed multifaceted commitments to managing urban resources and protecting urban environments. On the East Coast, New York City, Boston, Portland, Maine, and others have followed suit. The Council of Mayors of the United States is very active on issues of urban sustainability.

In Europe, many countries have adopted principles and drafted plans that target the pressing issues of the regions. Concurrently, municipal governments and private organizations are also developing protocols for assessing and promoting urban sustainability. One example of this is the European Green City Index. This index is the product of an effort by Siemens Corporation to provide an assessment metric to be used on any city. Figure 6.4 places a number of European cities in terms of the index and a measure of economic output.

Today, we can understand the emergence of green city plans and initiatives in a broader context of diverse efforts from around the world. The purpose here is not

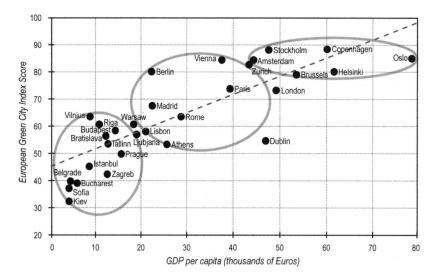

Figure 6.4
Siemens European Green City Index: a rough grouping shows that eastern European cities score low both in terms of the index and GDP per capita; the middle consists of western European cities with temperate climates, and those scoring highest are to found in the northern latitudes. Of interest here is the fairly strong correlation between the index and both per capita economic output and climate.

to provide a simple listing, as the reader is likely well aware of many of these efforts, but to classify them in terms of their dominant priorities and essential characteristics.

Sustainable City and Neighborhood Guides and Organizations

The diversity and global reach of the problems associated with contemporary urbanization has led to the founding of numerous extra and nongovernmental organizations focused on the challenge. These third-party groups have played key roles in introducing and educating many cities to the process of formulating goals and articulating plans that are globally significant and locally practical.

Green city plans defined by third parties are, at their initial conception, more generalized than a locally formulated plan or set of initiatives. This is a necessary result of the development of plans that start from a set of common generalized priorities and principles, such as response to global climate change, and progress toward addressing specific local conditions and interests. Typically these kinds of plans are initiated by, maintained, and promoted by nongovernmental nonprofit organizations that take as a central mandate to ensure the quality and comprehensive nature of the set of urban sustainability priorities. Therefore, these plans take on a systematic character by way of the fact that the criteria for action and success are generalized and applied to a variety of urban situations.

The number of organizations involved in this kind of work has increased dramatically since 2000. Today, the International Council for Local Environmental Initiatives, World Resources Institute, United States Green Building Council, Global Footprint Network, and many others are working with cities on strategies toward urban sustainability, the development of green city plans, community-led initiatives for healthier and liveable cities, and many other facets of the economic, social, cultural, and resource well-being of their cities.

Emerging Evidence of Success ... or Not

Generally, it is still too early to understand the effects of green city policies in a comprehensive way. While certain successes are clear, such as those of Curitiba and Portland, Oregon, the question remains whether policy initiatives have been able to substantially stem the advance toward an ever more resource-intensive world.

A global assessment arrives at a rather sobering state of affairs; ever more so because of the strength of the trend away from overall urban resource consumption.

There is considerable reason to question the actual results of green city plans, especially those new cities that have received so much publicity. In particular, two of these projects demand closer scrutiny: the Dongtan eco-city in China and the Masdar project in the United Arab Emirates.

Dongtan is far behind schedule and has been hobbled by a combination of local political dysfunction and a lack of process and approvals for developing urban infrastructure outside of the detailed and rigid strictures of the national government. Despite very high-level commitment that included an agreement between the president of China, Hu Jin-tao, and then prime minister Tony Blair, the international engineering firm ARUP of the UK, and the Shanghai Industrial Investment Corporation, the project to stem the enormous tide of development adjacent to Shanghai and replace it with a resource-efficient new eco-city model seems to have run out of momentum.

Masdar, in contrast, is running under full commitment and strong momentum, with support from the government of the United Arab Emirates and a cadre of international consultants. While good information regarding the actual status of the project is difficult to obtain, it is clear that there has been no lack of commitment from organizations within the UAE and no relaxing of interest of consulting groups from around the world.

Masdar has the challenge of delivering on its assertion to be the world's first carbon-neutral city. Clearly, this assertion comes by way of the very careful and strategic delineation of a system boundary around accounting of carbon emissions.

Both of these initiatives are important milestones as evidence of the establishment of an international level of awareness and commitment to the goals of urban sustainability. As such, each project is also subject to the complexities, unintended consequences, serious delays and other distracting and derailing dynamics of an uncertain future that characterizes any large-scale international effort. The serious question to ask, and a difficult one to answer, is whether these efforts warrant the monumental expenditures of financial and natural capital within the fairly resource-constrained world of urban sustainability?

One approach to an answer are the seemingly more promising series of initiatives involving the greening of particular urban systems within *existing* cities that have clearly identified critical resource and environmental priorities. While these initiatives certainly lack a utopian and shiny-new future, they may be the first indications of the urban revolution that has potential to deliver real resource gains for a sustainable and humane urban future.

In fact, a conception of green cities as an element of a future notion of strong sustainability globally is not generally the measure of success for all of the cities listed thus far and of any of the green city policies examined in this chapter. Unfortunately, frequently cited examples of sustainable city initiatives today are deeply flawed—socially, environmentally, and sometimes economically and politically.

It may be time to acknowledge that the pathway toward urban sustainability is primarily dependent on the ability of existing cities, especially those that are at the epicenter of massive population growth in the coming decades, to adopt innovative and effective policies that narrowly target important local resource and environmental issues. This may be facilitated by adopting a life-cycle perspective in designing and planning urban systems, as this will include not only the construction costs but also maintenance, use, and end-of-life costs, and therefore, solutions that might be more expensive to build but offer more efficient resource use during its life time might prove to be more feasible both economically and environmentally. This kind of targeted adaptation toward radical and strong urban sustainability can address the most pressing issues of the day while laying the framework for working across urban economic sectors and integrating urban infrastructure toward a future in which the city behaves in a radically different way. Certainly, high-profile projects such as Dongtan and Masdar play an important role in advocacy for a distant future of near-zero urban living and working, but the need to deliver actual reductions in urban footprints today, or very soon at least, are too great.

This is why it is imperative that the major primary cities of the world and the enormous number of mid-sized cities begin a process of implementing the most promising urban sustainability measures that address their particular situation.

By delivering important results in the short term (reductions in urban pollution, including carbon emissions, increases in mass transport usage and building consumption efficiency, management of waste streams, for example), existing cities will begin to contribute to the global project of addressing the resource intensity of society. The approach to a sustainable urban future then becomes not a utopian leap toward a zero-carbon ideal, but a many-faceted and necessarily incremental accumulation of real gains most critical to the local urban situation. This is just another way to state that the best approach toward sustainability is to think globally but act locally—or think global impact while acting at the urban scale. Real results in the short term and in the local urban context will lead to momentum toward a more radical sense of global urban sustainability.

As described in this chapter, the current drive toward urban sustainability is neither strong enough nor aspiring toward strong sustainability. As such, it is possible to arrive at the conclusion that it is flawed if considered as the best effort that contemporary society is willing to make. However, if considered as a transition toward strong sustainability, current green city initiatives and the organizations that promote them are an important bridge to the future.

Yet, it is not apparent that we understand what the future of urban sustainability really entails. The "long way down" is considered alternatively as the latest pessimistic and historically naïve example of crying wolf at a time when we should be focusing all of our energy on revolutionary and creative breakthroughs. On the other hand, the long way down suggests, at least to some, the dawn of a new, more humane era of responsible consumption and innovative production leading global society toward an anthropogenic ecology of efficient, well-mannered, and techno-logically sophisticated cities.

Some believe we are collectively at the beginning of an era when urban agglom-erations are the gateway through which humanity travels toward innovation, afflu-ence, cultural diversity, and ultimately sustainability (Glaeser 2011a). Keeping in mind the faults of techno-obsessive utopian visions of future cities, these optimists can point to a generally more affluent, well-educated and possibly more secure urban population than ever before. More people than ever today live, work, and provide, albeit at slim margins, the basic staples for life. The massive starvations, pandemics, social upheavals and widespread violence envisioned by various writers and thinkers over the past fifty years have not generally materialized to the scale predicted. At least not yet.

7

Urban Typologies: Prospects and Indicators

As cities have adopted the priorities of resource efficiency, focus has shifted toward ways to monitor progress and define success. Accompanying the proliferation of green city initiatives described in chapter 6, there has also been a great deal of work to formulate useful metrics of urban sustainability. The result has been the formulation and compilation of urban sustainability indicators: specific quantitative and qualitative measures of every aspect of urban sustainability. Aside from the obvious need to track progress or the lack thereof, these efforts are also meant to establish a robust and scientifically sound basis upon which generic indices can be used in comparative analysis of the resource consumption of a range of cities. Enlightened metrics tend toward indices based on a holistic urban assessment that considers the prospects for reducing the environmental consequences of urban consumption while supporting economic, social, and cultural priorities. This chapter will briefly review types of indicators and contrast this approach with a typological framework for understanding the drivers for distinct levels of global urban resource consumption.

Indicating Success

Indications of progress are found along a gradient between a less-than-ideal state of affairs and a desired end. Engineering design often entails having a rather precise sense of the performance requirements of desired results as a precursor to establishing the measures for design success. Similarly, architectural design also stipulates a set of priorities for performance while extending its reach to include complex issues of cultural meaning and production, social norms and customs, ecological concerns, and other issues that can belie strict quantitative measures of success. Urban planning and design extends these concerns to include the full range of human activities supported by and contained within the city. Clearly, quantitative measures of success are wholly inappropriate and inadequate for assessing the health and life of a city in its every aspect. Even beginning to define the meaning of the success of a city often becomes impossible without an extensive review of diverse and incommensurable matters spanning a vast range of human experience.

Considering the environmental and ecological consequences of urbanization while simultaneously accounting for the resource demands of urban activities is a huge undertaking riddled with fundamental methodological questions and plagued by a lack of reliable information and high-quality data. However, reaching conclusions about progress toward improving the urban environment while reducing the resource intensity of urban consumption is important; the need is clear and the benefits could be extraordinarily consequential.

Along the trajectory from rural to urban conditions, the prospects for sustainable society depend greatly on the nature and volume of resource consumption. A settlement within a rural context has the prospect for achieving respiration based almost exclusively upon a solar economy. In fact, pre–fossil fuel rural settlements had no other option. The total annual material flow for these agrarian settlements amounted to less than one ton of materials per capita (including fuel for energy). Material throughput per capita in many highly urbanized developing nations reaches upward of 50 tons per capita and above. Yet the agrarian scenario is not a thing of the past, as 1.4 billion people live today without reliable electricity (World Wildlife Fund 2011).

Urban Sustainability Indicators

An indicator establishes a measure of, and communicates progress toward, a specific goal. The specific purpose of any single indicator is dependent on the scope of its application and its role in assessing data in a useful way, communicating a specific intent, and supporting the decision-making process of a well-defined individual policy maker or group of stakeholders (Olalla-Tárraga 2006). A collective set of indicators represents the entirety of priorities for a community or city, region, or nation. These collections are often the core of urban sustainable plans.

The mandate of sustainability indicators has emerged as holistic, comprehensive, and integrative, in contrast to the reductionist and compartmentalized nature of traditional indicators addressing separate economic sectors, social issues, or environmental stressors (Haberl et al. 2004). As calls for sustainable development have proliferated in cities of contrasting political and socioeconomic conditions, efforts to harmonize frameworks for urban sustainability indicators have given way to diverse approaches and perspectives. Despite this increasing diversity, common features are often found between contrasting plans. Most lists of sustainability indicators can be classified as belonging to one or another type, or a combination. Domain, issue-oriented, and sectoral sustainability indicators focus on specific elements of the economy, the environment, or social fabric of society. Causal and mechanistic indicators link a particular phenomenon (such as an urban heat island effect) to the causes and mechanisms that produce it. Integrative and aggregated indicators attempt to synthesize several complex dynamics for the purpose of highlighting the need for equally integrated policies (Olalla-Tárraga 2006). Additional classifications are offered in the literature, such as policy performance indices, resource-based indices, and many others (Organization for Economic Cooperation and Development 2001).

Urban sustainability indicators generally originate from four separate sources: (1) academic research groups, (2) urban sustainability organizations, (3) regional and national governments, and (4) cities themselves. The motivations behind the development of criteria differ for each organization in obvious ways. There have been many efforts since the 1970s to establish robust indicators for sustainable development, and the development continues today as the priorities and challenges for sustainable development evolve.

For the research organization there is the interest to provide a more robust scientific basis for assessing and tracking the various aspects of urban consumption and production. Research is active in the assessment of both physical and non-physical urban systems, as researchers tackle both the issues confronting pathways toward urban resource efficiency and urban livability.

Urban sustainability organizations develop indicators for the purpose of serving the very real need for immediate application and action. These organizations are both for-profit consultancies and not-for-profit organizations that count municipal policy makers and governments as their direct clients. For-profit consultancies are challenged with the task of providing practical, real-world guidance to their clients for the purpose of guiding decisions in real time. Not-for-profit organizations are also challenged in this way, though the intent of their work also includes providing a range of cities with frameworks of success that can be applied.

Regional and national governments, and international governmental collectives like the European Union, also develop urban sustainability indicators. Often these urban indicators are ensconced within priorities at a much larger scale and are considered important insofar as they may contribute to regional and national climate change and resource-efficiency goals.

Sustainability indicators developed for use by individual cities are similar to those produced as "generic" models, with the addition of important local issues. In fact, these municipal projects tend to closely follow the guidance of generic models. These issues result from local conditions such as climate, topography, biodiversity, industrial base, cultural and historical foundations, and many others. City policy makers who direct the development of sustainability indicators for their own use are highly influenced by the political, economic, and social context in which they are working. Significant differences in the cultural context of the city can significantly affect the strategies they adopt. In 2011 and 2012, a small but growing number of smaller communities in the United States have been withdrawing their membership in the International Council for Local Environmental Initiatives (ICLEI) and rejecting general principles of urban sustainability. While the adoption of urban sustainability priorities still overwhelms these rejections, the example simply indicates that local conditions are a significant consideration that influences the development of urban indicators.

Often a combination of these organizations will collaborate to adopt or newly develop indicators for their own purposes. The criteria developed by ICLEI may be customized to better address the priorities of a particular city. An individual mayor may insert major considerations into a generic model for any number of reasons. A consultancy may structure their indicator listing in ways that improve their position for business development and marketing. It is important to acknowledge that urban sustainability indicator listings, especially those adopted and used by cities, are just as prone to the pressures of local special interests, historically and culturally rooted preferences and concerns, and all of the diverse foibles and inconsistencies that are the hallmark of human organizations.

In any case, the following is intended to provide an overview of the basic structure of a number of typical urban sustainability indicator lists.

Newman (1999) provides a comprehensive listing of indicators for urban sustainability that includes:

- energy and air quality
- water, materials, and waste
- land, green spaces, and biodiversity
- transportation
- livability, human amenity, and health.

Contained within these items are most of the issues to be found in many indicator listings: pollutants, urban heat island effect, nutrient flows, eutrophication, hazardous materials, land use, industrial effluence, mass transit, urban pathologies, asthma, infant mortality, crime, housing, density, and many others. Resource-oriented sustainability indicators are most often structured in terms of individual resource flows—water, energy, materials, etc.—and the urban metabolism framework, oriented as it is on resource flows, is particularly responsive to resource-oriented sustainability indicators.

Indicator lists are as diverse as the set of topics they attempt to address. As such, efforts to standardize a conventional hierarchical listing are to be considered within the context of a diverse set of priorities and conditions. It may be inevitable that conventional listings will remain quite generic and therefore not useful for direct application to specific urban zones.

Customized city indicator lists often tend toward exhaustive listings, even while it is clear that many items are either difficult or impossible to address and a more hierarchical approach would serve the city better. Pressed for resources and attempting to address emerging issues in economically viable and politically palatable ways, the drafting and adoption of city plans is often enormously challenging. However, exemplary examples are not difficult to find.

Santa Monica, California, lists the following as indicators for their Sustainable City Plan:

- resource conservation indicators
- environmental and public health indicators
- transportation indicators
- economic development
- open space and land use
- housing
- community education and civic participation
- human dignity.

Material intensity per unit of service is one kind of measure that is particularly useful in the urban metabolism context. Measures of urban resource consumption such as vehicle miles (or kilometers) traveled per household hold particular promise for holistic assessment of urban consumption as related to specific elements of the urban infrastructure. Even measures of the infrastructure *resource content* per household or per capita can usefully inform decisions about the growth of cities.

Life-cycle assessment and other measures of resource consumption are described in detail in other parts of this book.

Urban Typologies

The characterization of urban form is a key element in the determination of prospects for urban sustainability. This is due to three main reasons. First, the nature and intensity of resource consumption along the gradient of urbanization is highly dependent on the coupling of population density and density of services and infrastructure, particularly transportation density. The relationship between overall numbers and densities of urban populations and the services that they access, connected by the transportation network, substantially determine a facet of overall energy and material consumption. A diverse mix of critical services (fuel, groceries, health care, and security) located proximate to densely populated areas served by a thick set of transportation arteries ensures that low levels of energy consumption in transport will be achieved.

Second, the placement, scale, and configuration of the urban built environment affect the manner and intensity with which households, commercial establishments, and industry consume energy and materials. Buildings that are sited for positive solar gain for heating and natural ventilation for cooling consume less energy. Buildings in northern climates with high volume-to-surface ratios are generally more energy efficient, requiring less energy for the smaller enclosed volume and minimizing the loss of heat from a minimized exterior surface area.

Third, the dynamic interactions between buildings, open and green spaces, and urban infrastructure provide clues about the concentration of heat and urban pollutants that give rise to unhealthy and energy intensive "hot-spots" in the city. The primary phenomenon is the urban heat island effect. It is now clear that the configuration of buildings and urban canyons, the disposition and location of building waste heat exhaust outlets, and the surface materials of streets, sidewalks, green areas, building facades, and roofs all contribute to the thermodynamics and airflow of urban spaces.

While the prospects for urban sustainability are substantially dependent on the characteristics of urban form, as explained by the three aforementioned reasons, achieving substantial progress involves a variety of challenges. For example, the planning, design, and maintenance of transportation networks is most often conducted by regional authorities, while regulation of the design and construction of buildings is often the responsibility of a more local authority (Wheeler 2000). Coordinating the density of one versus the configuration of another is not undertaken under common goals by closely associated regulatory and governmental agencies. Achieving real progress toward any kind of predetermined optimal combination of population, service, and transportation densities will require an unprecedented level of municipal and regional authority cooperation.

Also, the density and activity contained within sections of urban settlements are loosely determined by upper bounds of floor area ratios, site coverage percentages, building occupancy types (residential, commercial, industrial, for example), and sometimes building height maxima and massing limits. These parameters of the built environment are established incrementally and define a relatively monolithic framework within which the ever-changing economic development and real estate market ebb and flow.

The complexity of these relationships can best be approached through an assessment of the dynamics of the systems that are adjacent to one another, interacting at various levels and evolving over time.

When it comes to complex systems, the overall measure of success can become quite complicated. An instance in which one requirement can only be met by compromising the delivery of another necessitates balancing these two and deciding what combination is acceptable. When various requirements are related to one another in complicated ways, the exercise to optimize can become quite complicated. Some situations processes cannot be concluded without a multi-objective optimization that requires a decision to be made between one set of possible outcomes and another. In the end, it is necessary to make difficult decisions based on a balance of results deemed to be acceptable, though not ideal.

Considered through this lens, the determination of progress toward sustainability can seem daunting. Urban sustainability is no less challenging. The broad issues are well known. Urban sustainability targets the health of the environment within the city and the surrounding region as well as globally. It also addresses the economy as an engine for urban growth, provisioner of services and welfare for the urban populace, and source of employment for residents. The economic context within which the city is situated, its country and international economic network, are also the concern of indicators of urban health. Finally, a sustainable city is a humane one that provides a safe and fulfilling cultural, educational, and social setting for living one's life. It provides a healthy and just context for urban residents. The combination of these is a measure of the "well-being" of the urban center.

For example, the city of Melbourne has developed a reporting system that includes indicators under the following headings:

- a human city
- a sustainable city
- a prosperous city
- an innovative city, and
- an efficient and effectively managed city.

Sustainability indicators are directed at each one of these areas of urban well-being. Indicators that reliably and accurately reflect the "health" of every one of these elements of the urban context have been the subject of intense efforts at formulation and refinement for several decades now. In many respects, several systems for measuring urban sustainability are now fully operational and informing effective policy making at the regional and municipal levels. Organizations such as the International Council for Local Environmental Initiatives, Leadership in Energy and Environmental Design, and others have been fully invested in developing indicators that reflect real returns in resource efficiency while remaining accessible and practical to local, regional, and national government policy makers, community organizations, and others. Chapters 1–3 discuss these efforts and elaborate on the process for implementing and engaging with the latest indicators to supply useful information to policy makers.

Urban indicators have been a part of the policy instruments launched as a result of large international conferences that have addressed the urban future. Among others, the 1972 Stockholm Conference on the Human Environment, the 1992 Rio

Earth Summit (United Nations Conference on Environment and Development), the 1996 Istanbul Conference on Human Settlements, the 2000 Millennium Development Goals, and the 2002 Johannesburg World Summit on Sustainable Development are a few of the gatherings that have led to the formulation of policies that depend on useful indicators.

In 2005, United Nations Secretary-General Kofi Annan spoke about the need for a focus on the urban future. In his remarks he pointed out that "the rapidly increasing proportion of people who are living in urban areas—more than 60 percent by 2030 ... presents profound challenges, from poverty and unemployment to crime and drug addiction, ... and in too many of the world's expanding towns and cities, environmental safeguards are few and planning is haphazard." This international acknowledgment of the critical need to address the urban future requires measures that indicate to diverse parties whether progress is being made. This is the role of urban sustainability indicators.

However, fundamental questions regarding the efficacy of urban sustainability indicators remain. To date, very few important municipal decisions have been based substantially on indicators that relate urban production or consumption to the required resources for those activities. Resource-intensity indicators, such as material intensity per unit of service metrics, while represented in various forms, still do not provide the kind of leverage necessary for policy makers to make decisions about the allocation of municipal resources.

The larger of context of indicators begs the question of the prospects for urban sustainability. What are the ultimate possibilities for engendering a strong urban sustainability? Are there limits to urban resource efficiency?

A simple answer would be yes, in the sense that every individual—urban or not—requires a minimum daily input of food, air, and water while being protected from the uncertainties of the environment. This makes it very clear that individuals living in challenging climates—particularly the northern and far southern cold, hot, and arid climates—require additional resources to be protected from the uncertainties and extremes of these environments.

Much of the intellectual effort to build the capacity for a sustainable urban world is focused on developed regions. As described in chapter 11, this belies the fact of the locus of massive urbanization residing in the developing regions of the world.

However, the role of the developed world may best be characterized as the catalyst for driving global sustainability both with technological innovation and socially and policy-oriented breakthroughs. Sadly, the most prevalent form of urban sustainability can be found in the sprawling cities of the developing world. Clearly, this version of a sustainable urban future is not a future to aspire to. Rife with problematic health and environmental conditions, challenged by a lack of critical water and energy resources and sustaining not a vibrant, just, and humane urban society but one of rampant crimes against the weak, especially women and children, lack of proper education and employment opportunities, this version of the future is a sobering call for alternatives.

Yet the prospects for reaching a humane and sustainable urban future in the developing world are hobbled, ironically, by the emphasis on resources above all. Indicators most relevant to cities in the developing world are slightly different than those in the industrial world. Crime, corruption, governance, health, and employment are key elements of indicators for developing region cities, but the most important indicator focuses on poverty.

Poverty is the single most disruptive element toward the goal to achieve a humane city. It is also understood as holding a higher priority than even environmental issues, though the two are often symbiotically related. In developing countries and regions, cities are host to enormous populations living under, and often far under, the official poverty line.

A Global City Typology

One major challenge of urban sustainability will be the development and implementation of policies that target the most critical elements of a sustainable future for particular cities. Today, many cities from Copenhagen to Singapore have taken on the challenge of defining their priorities for establishing a path toward their own sustainable future. Working locally, many city sustainability plans are precisely targeted on the most pressing issues of the particular urban space.

It stands to reason that pathways toward urban sustainability are highly dependent on the local situation and preconditions. While there are clearly a variety of issues that are common to all cities—transportation, energy for households, water, materials acquisition and distribution, waste—different types of cities beg the question of whether or not there is a need for distinctly different types of strategies to achieve the best pathway toward sustainability for different "types" of cities. Grouping cities in terms of essential attributes possibly allows a better understanding of the nature of urban resource consumption. Linking the nature and intensity of resource consumption to essential attributes like population, spatial size, location, or the size and intensity of the urban economy may provide a better sense of the ways in which urban areas exact a toll on local and global resources. In addition, initiatives that attempt to improve the efficiency with which a city utilizes critical resources have been undertaken without a clear understanding of the general scope of resource consumption in cities worldwide. How do cities differ in their resource consumption? How are differences manifest and what causes differences in resource consumption?

Having a better idea of the intensity and nature of the flow, consumption, and transformation of resources and the generation and dissipation of urban waste across a broad collection of the world's cities may allow for better decision making across national boundaries. Approaching these differences in a constructive way has led to the development of a city typology, or a grouping of cities based on a limited set of similar attributes. This section describes the development of this typology.

The Development Methodology of the Typology
The typology presented here relates a set of widely available urban data to distinct levels of aggregate urban consumption of specific resources. The goal of the work has been to show that there is a direct relationship between a limited and specific set of urban facts and the associated nature and intensity of urban consumption. Structuring this relationship has led to the beginnings of a global typology of urban resource consumption (Saldivar-Sali 2010).

The first step in achieving a useful urban typology is clarity of the purpose for grouping, or *clustering*, cities. In this work, the clustering achieves the purpose of establishing a broadly comparative set of urban resource-consumption intensities that serve as a framework for categorizing the resource intensity of individual cities. The framework only defines generic categories of urban consumption as low, medium, and high, and places cities in groupings in which the likeness between them is much greater than their likeness to cities placed in other typological groups. In each type group described below, member cities display remarkably similar attributes of urban resource intensity that warrant the formulation of this typological framework.

The second step taken in the development of this framework was the demonstration of a predictive capacity inherent in the methodology used for establishing the typology. Simply put, a framework that groups like cities, for which much is known, may be useful in providing guidance to assessing the nature and intensity of resource consumption of cities for which little data is available. This predictive capacity is a direct result of the methodology used to cluster cities in the framework presented below. To capture a predictive capacity, the method of classification trees was employed here. Building and training classification trees with comprehensive and high-quality consumption data provides a model that serves as a first-order predictive tool for assessing data-poor cities (Breiman et al. 1984; Provost, Fawcett, and Kohavi 1998; Provost and Fawcett 2001; Perlich, Provost, and Simonoff 2004; Loh 2009; Saldivar-Sali 2010).

Placing cities within a spectrum of low to high overall resource consumption may seem like a relatively easy task. Certainly, the typology presented below does not disappoint in illustrating some obvious comparative relationships. Cities in developing regions are generally less resource-intensive than those in developed regions.

Cities of high affluence and very low densities display some of the highest rates of overall resource consumption. Cities in the tropics consume less energy in the built environment than those in much less hospitable climates in northern latitudes. However, these obvious conclusions are not particularly useful in revealing aspects of urban metabolism that may be linked to more complex combinations of urban attributes. Also, these simple observations do not allow us to extend our understanding of the urban metabolism of cities possessing combinations of attributes that do not clearly indicate placement within the spectrum from mild to intense urban resource consumption.

Independent and Dependent Variables of Urban Resource Consumption

Four independent variables were identified as predictor attributes of city resource consumption: affluence, population, population density, and climate. Overall population and climate are clearly influential attributes in the resource consumption of cities. For example, the energy-use intensity of buildings is significantly affected by the number of days during the year in which cooling or heating is consumed for human comfort. The absolute size of the population is also an obvious indicator of overall resource consumption; all else remaining constant, the more people consuming resources, the greater the overall resource consumption. It is also clear that the level of consumption for certain resources rises with growing affluence; for example, electricity, and in some cases water. Finally, population density has been shown to affect urban resource consumption in predictable ways that allow us to include this attribute in the set of independent, predictor variables (Krausmann 2008). Information about these four attributes is readily available for almost every city on earth.

This typology characterized the intensity of urban metabolism through an assessment of the consumption of eight resource types: total energy, total materials, electricity, water, fossil fuels, industrial minerals and ores, and construction minerals and biomass (the last four as specified by Eurostat in their material flow analysis standards). In the methodology employed here, these eight resources are dependent on the independent, predictor variables of affluence, population, population density, and climate (see figure 7.1). In addition, aggregate carbon dioxide emission was also used as a dependent variable. Carbon dioxide emissions act essentially as an overall consumption proxy most closely linked to energy consumption.

The classification trees that were developed assigned cities into groups as they displayed like consumption levels for specific combinations of independent variables. In this manner, a typology was constructed that allows for a comparison between many cities. Possibly more important, the typology prompts an informed discourse regarding the actual causal mechanisms that determine the various levels of urban resource consumption found here and their relation to macroscopic urban attributes.

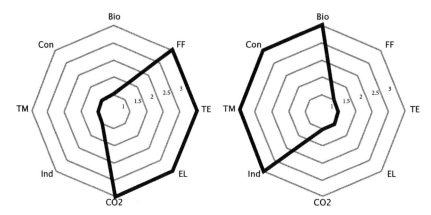

Figure 7.1
Two graphic representations of city resource-consumption types showing an energy-intensive city on the left and a material-intensive city on the right (Bio, biomass; FF, fossil fuels; TE, total energy; EL, electricity; CO_2, carbon dioxide emissions; Ind, industrial minerals; TM, total materials; Con, construction minerals).

In arriving at this methodology and its results, Saldivar-Sali (2010) cautions that this typology is a first-order, preliminary classification finding. The challenge of this grand project of classifying cities at the global scale is the paucity and low quality of available data. However, some of the results are insightful in their capacity to indicate complex relationships between specific urban attributes and the expected level of resource consumption.

For example, on the higher end of urban affluence, the typology demonstrates a significant range of urban resource consumption. Some cities in developed regions are much less resource-intensive in certain resources than cities at a very similar development level. Also, certain types of cities have been very successful at lowering the level of intensity of a specific resource. For example, Japanese cities have a small energy footprint, while still consuming electricity at high levels, an indicator of a good mix of non–fossil fuel energy carriers in a context of high affluence.

The results of the classification tree method are described below (see figure 7.2). Fifteen distinct city groups were derived using the framework regulated by a scale of low, medium and high resource consumption. Each type is described in some detail (Saldivar-Sali 2010).

Type 1

Cities in this category demonstrate low levels of resource consumption in all eight categories, except for water, which is found to be low to medium. The cities found

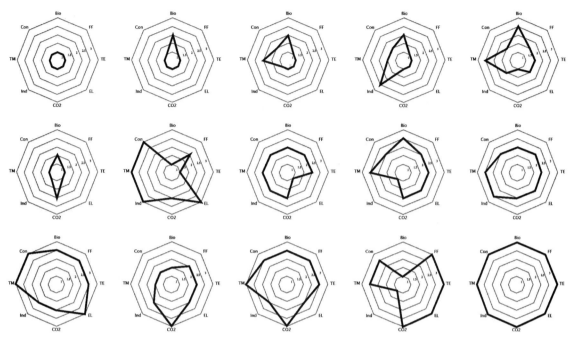

Figure 7.2
City types: The types are numbered beginning at the far left top row and proceeding to the right and down, beginning again at the left hand side of each row. Type 1 is found the upper left-hand corner, and type 15, the lower right-hand corner.

in this category are located in the most challenging developing regions: cities in India (Kolkata and Naihati), Indonesia (Jakarta and Surabaya), Cambodia (Phnom Penh), the Democratic Republic of Congo (Kinshasa), Freetown (Sierra Leone), and Sana'a (Yemen) are among those found in this type.

Type 2
As with the first type, type 2 cities demonstrate low levels of resource consumption in all eight categories, except for biomass, which is found to be medium, indicating the dominance of an agricultural economy. Cities in this category include Lagos (Nigeria), Addis Ababa (Ethiopia), Dakar (Senegal), Guatemala City (Republic of Guatemala), Bamako (Mali), Nairobi (Kenya), Mumbai (India), Quito (Ecuador), and Yangon (Myanmar), among others.

Type 3
Type 3 cities present a combination of low and medium resource intensity, with low consumption of total energy, electricity, fossil fuels, and industrial minerals.

Emission of carbon dioxide is also low. Consumption of construction minerals is at low to medium levels while biomass and total material consumption are at a medium level, and water consumption is found to be medium-high. Type 3 cities are largely agricultural with a growing industrial base. Consumption in construction minerals reflects the expanding infrastructure. Cities of this type include Cali (Colombia), Panama City (Panama), Manila (Phillippines), and San Salvador (Republic of San Salvador).

Type 4

Type 4 displays low total energy, fossil fuel, and electricity consumption, along with low carbon dioxide emissions. Construction minerals are consumed at a low to medium level, while biomass and water consumption are at medium levels and industrial minerals and ores at medium to high levels. Urban economies within this type are found in developing countries fueled by significant biomass-based industries. However, the presence of abundant mineral resources also drive both mining and light manufacturing. Cities of this type include Lima (Peru), Tunis (Tunisia), and Rabat (Morocco).

Type 5

Type 5 present evidence that carbon emissions are low while total energy, fossil fuels, electricity, industrial minerals, and ores and construction minerals display a low-medium level and biomass and total material consumption are high. This city type is also dominated by industries fueled by biomass, though carbon emissions are kept low through significant hydroelectric power resources. Cities here include Montevideo (Uruguay), Durban (South Africa), and Curitiba (Brazil).

Type 6

Total energy, fossil fuels, electricity, material consumption, construction minerals, and industrial minerals and ores are all low for type 6 cities. Biomass and water are found to be low-medium and carbon emissions are medium. Cities of this type show evidence of early stages of industrialization based on carbon-rich energy fueled by coal and oil. Several Indian cities are represented here, including Delhi, Bangalore, Hyderabad, and Chennai, as well as Ho Chi Minh City (Vietnam), and Cairo (Egypt).

Type 7

The cities of type 7 consume biomass and total energy at low levels, fossil fuels at a medium level, emit carbon dioxide at a medium level, and consume industrial minerals and ores, construction minerals, and total materials at high levels. Electricity consumption is high. This type is narrowly delineated almost exclusively by

Japanese cities because of the particular combination of a lack of large agricultural firms (due to the price of land and the mountainous terrain), a well-developed industrial base, and an advanced energy-efficient economy. While total energy is at a low level, a high level of consumption for electricity indicates an affluent standard of living. The particular combination found in this type may also be an indicator for an affluent, industrialized island economy.

Type 8
Electricity consumption is low and everything else jumps to medium for type 8. In this type, electricity consumption is a good indicator of a relatively low standard of living, despite the consumption of all other resources at a medium level. This combination indicates significantly industrializing economies with abundant natural resources, or ease of access to these resources. Several of the countries hosting cities of this type are home to large cement manufacturers. Included here are Beijing and Shenzhen (China), Brasilia (Brazil), Mexico City (Mexico), and Istanbul (Turkey).

Type 9
Within type 9 only industrial minerals and ores are at a low level of consumption. Total energy, fossil fuels, electricity, and construction minerals are consumed at medium levels and carbon emitted at a medium level. Biomass reaches a level of medium-high. Cities of this type are found within *transitional economies*. In assessing membership within this type, Saldivar-Sali (2010) expands the notion of transition economies to include those that are emerging from controlled, socialist, or communist economic structures as well as those making the transition from heavily agricultural and heavy industry toward more diverse economic development. Countries that are considered members here are several countries of eastern Europe as well as Iran, Libya, Portugal, and Argentina, for example. Cities include Belgrade (Serbia), Tripoli (Libya), Buenos Aires (Argentina), Tehran (Iran), and Lisbon (Portugal).

Type 10
For type 10, all energy consumption, construction minerals, and carbon emissions are found at the medium level. Industrial minerals and ores, total materials, and water are found at medium to high levels. Type 10 cities are found in developed countries with diverse and industrialized national economies. Many of the cities of this type are dense European cities with strong service-based, non-material-intensive business sectors. Consumption as a factor of significant affluence is tempered by the density of the actual urban form and associated efficiencies gained through well-developed mass-transit systems and concentrated population densities. Cities include London (UK), Berlin (Germany), Dublin (Ireland), Rome and Milan (Italy),

and Madrid and Barcelona (Spain). Also included in this type are cities of advanced economic transition, including Kiev (Ukraine), Santiago (Chile), Sofia (Bulgaria), and Caracas (Venezuela), for example.

Type 11

Medium consumption is observed for carbon and industrial minerals and ores consumption; medium to high for water, biomass, total energy and fossil fuels; and high for electricity, total materials, and construction minerals. Cities of this type extend the consumption of energy for type 10 incrementally upward, reflecting lower density and higher affluence; Paris (France) and Dubai (United Arab Emirates) are examples.

Type 12

Total materials, biomass, and construction are consumed at a low-medium level; total energy, electricity, fossil fuels and industrial minerals and ores at a medium level; water at a medium to high level; and carbon dioxide at a high level. Mining activities and coal-burning dominate the economies of several of the countries with cities in this group. High carbon emissions indicate this, even though electricity consumption remains relatively lower, indicating lower affluence than either type 10 and 11. Cities include Shanghai (China), Tel Aviv (Israel), and St. Petersburg (Russia).

Type 13

The lowest level of consumption, at medium, is with industrial minerals and ores, while both total materials and carbon emissions are at a high level. Everything else is found to manifest consumption at a medium to high level. Cities found here are located in industrially advanced and developed countries that contain the largest producers of coal, cement, food and beverages, textiles, and agricultural and cellulose products. Cities include New York, Los Angeles, and Seattle (USA), Helsinki (Finland), and Copenhagen (Denmark).

Type 14

Industrial minerals and biomass are consumed at low levels in type 14. Then there is a jump to medium-high with total materials, water and construction minerals, and high for all energy components as well as carbon dioxide. Type 14 is reserved for cities demonstrating a very particular mix of urban respiration that includes a net export of industrial minerals, very low biomass consumption due to conditions of an arid climate, strong consumption of construction minerals indicating large-scale urban construction, strong consumption of water some of which is satisfied with imported water, and a high level for all energy indicators, including carbon emissions. Cities with this particular urban metabolism profile are all found in petroleum producing nations of the Middle East. Cities include Riyadh (Saudi Arabia), Abu Dhabi (United Arab Emirates), Kuwait City (Kuwait), and Doha (Qatar).

Type 15
At the extreme end of the spectrum for urban resource intensity, all eight resources are consumed at high levels, and carbon dioxide emission is high as well. These cities are high consumers for all of the obvious reasons: low densities requiring significant energy expenditures by automobile are combined with high affluence, while challenging climates require significant building heating and cooling energy expenditures. Cities include Phoenix and Chicago (USA), Toronto and Montreal (Canada), and Sydney and Melbourne (Australia).

These results are the first indication of the possibility of characteristic urban resource profiles at a global scale. For the most part, teasing out causal links between certain attributes of cities and their representative resource intensities can only be done in a preliminary, tentative way. Some links are clear; dominant industries affect the nature and intensity of resource consumption, while affluence affects consumption but does not necessarily mean high-intensity consumption of all resources. The indigenous natural resource base, especially with regard to ores and fossil fuel carriers, can significantly alter the resource profile in narrow ways. Understanding these linkages and finding the causal mechanisms that underlie them is the next step for this typological work.

The Future of Urban Sustainability Indicators

To speculate about the future of urban resource indicators without grounding in the larger context of national and international action addressing global resources would be to risk becoming mired in an analysis of local politics and the minutiae of regional regulatory frameworks. Therefore, this assessment takes cues from the international context.

In this light, it is clear that the near future of the debate on urban sustainability indicators will include questions of equity and social and environmental justice. Just as the debate on global warming has introduced the idea that high-income households, irrespective of their country of residence, have a larger responsibility for reducing carbon emissions, the debate regarding sustainable cities should recognize the dire needs of many of the developing world's urban poor. Chapter 11 will highlight the particular challenges of cities to be found in the non-OECD developing world.

Much of the discussion of global climate change, allowable carbon emissions per capita, and the need for another round of international agreements on the topic has focused on comparisons between nations. At the United Nations Framework Convention on Climate Change, it was agreed that the responsibility for reducing carbon emissions rested with developed nations. However, as described elsewhere in this book, the acceleration of urbanization already outpaces that of the developed world and almost all of the coming urbanization of the next few decades will occur in developing regions. With the inevitable rise in affluence and more intense resource consumption that is the hallmark of urban living, there is a real need for developing countries to participate in low stabilization targets for future carbon emission.

An intriguing new pathway for the discussion has been unveiled that highlights the contribution to consumption of the wealthiest population globally. The idea of "common but differentiated responsibilities" identifies the level of carbon emissions of individuals, not nations, as the more productive topic for discussion (Chakravarty et al. 2009). High-wealth households are essentially high carbon emitters whether they live in India or the United States.

Taking this perspective into account, it is clear that the prospects for sustainability for cities in developing regions are limited. That is, if one assumes an increase in development of these regions coupled with the increase in affluence that is borne out by urban living, reductions from developing regions are not only not unlikely but not the best choice in balancing the need to address poverty, crime, and environmental conditions.

8

Complexity and Dynamics of Urban Systems

This chapter will review the characteristics of complex urban systems and briefly introduce the fundamentals of system dynamics for the purpose of illustrating productive applications of the conceptual framework and methodological process in the assessment of urban resource-consumption behavior. The approach, while dependent on the individual decisions of members of households and directors of companies, will be taken at a macroscopic level—integrating the effect that policies and technologies have on a variety of related, but unlike elements of the urban context.

Urban Complexity

Cities are, and have always been, the most complex artifacts of human culture and civilization. No other single human artifact embodies as well the extraordinary diversity and systemic complexity that is representative of organized society. Encapsulated within urban space and time are innumerable instances of human relations—economic, cultural, and social—that drive our collection and consumption of resources toward constructing the built environment, urban infrastructure, and industrial capacity of society. Approaching an understanding of the city requires acknowledging an urban complexity that is both a unique anthropogenic construct and simultaneously evocative of the most complex biological organisms.

"Cities happen to be problems in organized complexity, like the life sciences" (Jacobs 1961).

Cities are the most complex human artifacts primarily because they concentrate and serve to facilitate the myriad range of human relations within buildings, on streets, in parks, every day and every year since the adoption of agriculture and the settlement of the first village. Complexity of transactions, whether verbal, physical, economic, social, and every other kind, is facilitated by the civic framework of private firms and public institutions, markets, government, cultural venues, and many other spaces and facilities. It is the nature and content of urban transactions that serve as the focus of perspectives on the resource intensity of cities. The flow of units, energy, materials, products, services, information, people, biodiversity, and so on defines the nature of urban space and delineates the measures and assessments of resource intensity and urban sustainability. As discussed elsewhere in this book, these inter- and intra-urban flows of resources are collected from the global hinterland and concentrated on the urban activities of businesses and people. The intensity with which these flows are mobilized and consumed substantially determines our approach to notions of urban sustainability, and while cities have never been sustainable within their boundaries they have always sustained the most complex and rich human experiences possible.

All of these urban human relations are knit together by the forces of mutually beneficial relations between individuals, groups of individuals, organizations, businesses, institutions, and governing secular and religious bodies. Central to the transactional glue of cities are the dynamics of agglomeration economics—balanced between the value-generating centripetal forces between workers and businesses that converge on the central urban space and the centrifugal forces of higher housing and land prices within that same space. These dynamics are readily apparent in central Singapore as in so many like and unlike cities (see figure 8.1).

Figure 8.1
View of Scotts Road, Singapore.

The forces of the urban economy are not disengaged from the physical reality of the urban place; its topography, climate, biogeochemical cycles. These forces necessarily act along vectors that function as conduits for the delivery and consumption of urban resources. These urban conduits are the large-scale elements of urban infrastructure and organization; the highways, electrical power grid, water system, land use designations, business concentrations, and so on that determine the path and efficiency with which urban resources culminate in serving all urban activity. For example, transportation plays a fundamental role between firms and workers in the agglomeration of urban economic transactions in the condensed space that leads to the founding and growth of a city. The reduced costs of urban transportation catalyze and support benefits accrued by both residents and firms. Simultaneously, the physical form and extent of urban transportation plays a fundamental role in determining the resource intensity of the provision of urban transportation. The actual physical dimensions, density, configuration, and extent of the transportation network and all of its elements substantially determine the resource intensity with which the provision of urban transport is delivered to residents, businesses, government, and institutions.

In addition, urban space creates the essential elements that foster the unique alchemy of innovation—that diverse blend of intellectual, social, technological, and logistical dynamics that give rise to the flowering of human knowledge, the catalyzing of invention, and the promotion of cultural, technological, and business achievement. Much important recent work has focused on this particularly urban product (Glaeser 2011b, 2011c). Considering the value of this aspect of cities, and incorporating this value into the discussion of sustainability, offers a critical dimension to the evolution of the notion of urban sustainability.

This may be done by linking patterns of growth to resource intensity of those urban settlements. Determining the effect on ecological functions, biogeochemical cycles, and other aspects of the urban environment as a result of urbanization along a gradient from center-city to rural delves into the actual geometry of urban form.

It is clear that the presence of urban complexity in all its aspects is to be found in every culture, location, climate, economic profile, and industrial mix that characterizes any particular city. Every city embodies complexity. From ancient Aleppo to contemporary Dubai, the urban context always plays a primary role as host for the diversity of transactions that characterize human experience. That is, urban complexity is independent of a vast array of distinguishing characteristics, both in time and space, that influence other aspects of a city from its transportation network configuration to its economy. Complexity is inherent in the very fabric of the urban context—it is the essential nature of cities.

Studies in urban complexity have often approached the study of cities quite independently of many other disciplines examining the urban world. This has been a source of strength as well as weakness in the accumulated results from complexity studies. Criticism has centered on the doubts of urban specialists that consider an emphasis on a geometric approach to urban complexity as too "simple." Models that reproduce nonlinear dynamics and employ techniques such as cellular automata, multiagent models, fractal growth, neural networks, rule-based urban decision making, and evolutionary models have been routinely applied to the urban space for some time now (Pumain 1998) and several reviews of the nature and content of this work have been written (Mulligan 1984; Bertuglia and La Bella 1991; Batty 1992; Wegener 1994). While it is true that these methods focus on specific attributes of urban patterns, complexity studies have been particularly effective in deriving a better understanding of patterns underlying mountains of urban data that represent the decisions, and various consequences of these decisions, that urban individuals make every day. Decisions about the path of travel for the daily commute to work and back, purchasing decisions, residential location, and employment decisions are just some examples of the range and complexity of decision making that can be further studied to inform the effort toward sustainable cities.

Types of Urban Complexity

Thus far, one type of urban complexity has been emphasized; that is, the complexity that is a result of urban transactions. This particular activity of urban reality can be considered the essential element, the catalyzing force, behind all urban space. Transactions between individuals of diverse skills as an essential urban dynamic have been cited since the earliest discussions of the origins of the city and complex economies.

"The origin of the city, then," said I, "in my opinion, is to be found in the fact that we do not severally suffice for our own needs, but each of us lacks many things." ... "As a result of this, then, one man calling in another for one service and another for another, we, being in need of many things, gather many into one place of abode as associates and helpers, and to this dwelling together we give the name city or state, do we not?" "By all means." "And between one man and another there is an interchange of giving, if it so happens, and taking, because each supposes this to be better for himself." "Certainly." "Come, then, let us create a city from the beginning, in our theory. Its real creator, as it appears, will be our needs." (Plato 380 B.C./2003)

Alongside the urban transaction, other aspects of urban reality require listing and brief description:

- urban spatial attributes under growth and shrinkage and shifting
- temporal dynamics at diverse scales, from real time to decadal and beyond
- urban resources, material and energy flows, and economic development
- large-scale physical systems and infrastructure, including ecosystem services

- organizational and institutional frameworks
- urban population and demographic trends
- political structures and governance
- cultural and social foundations.

Spatial dynamics and complexity is a central concern of the modeling efforts listed above and includes characteristics of the links of cities to their surrounding countryside and specific spatial attributes within the city fabric itself. In addition, definitions of the nature and spatial extent of urban places has complicated the study of cities as discrete complex systems as boundaries for the system are sought. The interaction between urban boundaries and the agents and resources that flow into and out of them are of central concern to urban metabolism studies. The flux of people and materials contributes to the temporal complexity of city form and behavior. For example, the role that commuters play and the designation of resource intensity due to large-scale transportation hubs, such as airports, are issues that also affect boundary definitions and assignment of consumption to one urban group or another.

In addition, the growth of cities is highly influenced by various evolutionary trends in national and regional economics, cultural and social movements, and political realities. The advent of information technologies not only affects the way that workers consider the "place" of work but also the emergence of urban innovation clusters. Political reality and changing funding priorities can directly affect demographic flows and urban resource needs. All of these are contributors, to some extent, to the physical attributes of urban form and the associated resource flows serving urban activities (Pumain 1998).

Urban Dynamics

Closely coupled to urban complexity are the dynamics that arise from the flux of physical things, like materials and energy, as well as nonphysical flows such as information, innovation, and financial instruments (see figure 8.2). While it is exceedingly obvious to state that the urban construct is inherently dynamic, achieving a practical understanding of those dynamics is not an easy task.

This chapter introduces systems dynamics as a way to "map" the elements that comprise a system and lead to the complexity of its behavior. An essential aspect of this mapping is time. Studying a system that acts in time necessarily requires a methodology that allows for a time series, or dynamic observation of behavior.

Another essential step in mapping system elements is the clear determination of the extent of the system under study. Clearly, a city exists within a larger regional, national, continental, and global context. Using system dynamics as a method of analysis requires the delineation of an urban boundary.

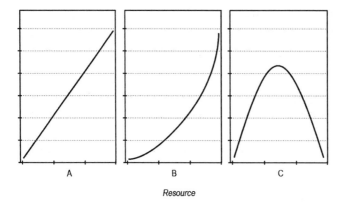

Resource

Figure 8.2
Generic behaviors linking growth in the urban economy with the increase in consumption of various resources. Material consumption increases linearly with economic expansion (A), exponentially for consumption of electricity (B), and reaches a maximum for urban water (C).

It also often requires the decision of whether to consider the interface between the urban system and the non-urban surroundings. There is now significant attention trained on the nature-human interface generally and, as a subset, the urban-nature interface (see chapter 3 for a discussion of urban ecology). Recent work has focused on the interaction between anthropogenic urban systems (such as civil infrastructure and transportation) and the natural ecologies and biogeochemical cycles within which the city operates. This multidisciplinary, multilevel approach seeks to model the primary interactions that determine the environmental effects of all manner of urban activity. Articulating the spatial, temporal extent of this interface and accounting for the interchange of materials and energy is an important aspect of understanding urban dynamics.

The essence of the interest in urban dynamics for advancing the field of urban metabolism is uncovering clear and essential linkages between dynamic urban trends and resource flows. How do changes in infrastructure design and engineering affect the regional biogeochemical cycles that contribute to passive urban resource flows such as water, air, particulates, even biodiversity? How does the changing nature of urban institutions and governance affect the ability of government and urban residents to increase their resource efficiency?

These questions are enormous and will require the attention of many researchers. However, some small prizes are emerging. For example, it is becoming clear that distinct resources display markedly distinct behaviors as related to trends in the urban economy as a whole. That is, the change in the intensity of consumption of different materials does not follow the same trajectory as urbanization progresses even as those resources are examined within one urban economy.

Urban Growth and Size

The complexity of cities and their behavior over time belie an interesting rule that seems to represent, if not govern their size and extent. The rank size rule, articulated by Zipf in 1949, states that the population of a city is proportional to the inverse of its regional rank. This can be mathematically represented as a power law with exponent of approximately –1 (Decker et al. 2007). Many studies have confirmed the presence of the rule that seems to imply the process of urban growth is scale-invariant, though the mechanism behind this invariance has been variously explained and vigorously debated. Other studies conclude that Zipf's law does not hold, that the data used to prove it are often incomplete and that there are significant deviations from it.

However, this is just one example of the intriguing and potentially powerful link between growth, size, and the flow of resources that serve these complex systems. In fact, one explanation for the consistency of Zipf's law is that human settlements—and society, by extension—are constrained in their extent by fundamental relationships between space and resource distribution. More work in this area may yield practical benchmarks that can be used in the development of more effective models for predicting growth and its associated resource intensity.

Complexity, growth patterns, networks, and size and scale are productively brought together dynamically through system theory. The remainder of this chapter invokes systems theory and system dynamics as a path toward better understanding urban behavior.

Systems Theory

This chapter has invoked the notion of a system without a proper definition. This is because, in terms of our understanding of cities, it is obvious that cities are composed of multiple interrelated elements that act together, in concert and in competition, within urban space and time. However, it is useful to denote the exact ways in which cities are systems and the utility of this notion for analysis and understanding.

A system is a bounded portion of the world. Within the boundary—the system boundary—we may find all manner of elements. If our system is a natural meadow, the system is composed of the animate and inanimate, organisms and minerals, interacting with one another in diverse ways. Depending on our interest, be it the biodiversity of the ecology of the meadow, the size of the population of field mice, or the concentration of insecticides within the spatial boundary of this particular meadow, we may draw the system boundary differently and track specific elements of this rather complex set of animate and inanimate elements. We could end up

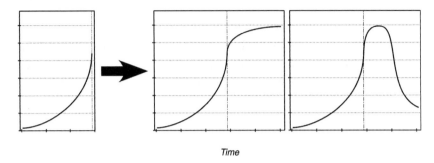

Time

Figure 8.3
A system boundary is also defined by the temporal scope of the key question. Limiting the temporal scope may lead to the premature conclusion of exponential growth when looking further reveals either a leveling off to a new plateau or a collapse and rapid decrease.

tracking a vast range of elements in our attempt to understand their interactions with other elements for our particular purposes. It is not likely that we would attempt to track *everything*. In other words, we are not likely interested in representing every physical thing and tracking every conceivable interaction between all elements. It is not likely that we would be interested in tracking the actual movement—every footfall—of every field mouse over the course of every day during its life. That is, we would likely not want to, or not *need* to, identify every element and every interaction between all elements to begin to satisfy our particular interest (Érdi 2008).

So, it becomes clear that the study boundary of any particular system is also limited (see figure 8.3). In fact, the boundary that is most useful to delineate in the study of any particular system is substantially determined by a key question: what is it about the system that one desires to know? Usually this can be articulated with a question; for example, what is the relationship between field mice and red-tailed hawks, or what is the concentration of insecticides over time?

The field that concentrates on the actual methodologies for understanding systems is referred to as *systems theory*. General systems theory was founded in the mid-twentieth century by a group of individuals from various disciplines, though the seminal book by the biologist Karl Ludwig von Bertalanffy formally introduced systems theory to the world (von Bertalanffy 1968) and heralded a proliferation of work based on its general methods. Diverse researchers from Margaret Mead and Donella Meadows to Howard T. and Eugene Odum adopted the general systems theory framework, extended its principles, and derived various original pathways for understanding the dynamics of complex systems.

In particular, a great deal of attention has been focused on the dynamic behavior of complex systems. Understanding interactions between elements within a system usually involves time—and often, also space. Understanding interactions by observing and explaining dynamic phenomena is central to systems theory. Therefore, much of the work in systems theory is called system dynamics.

In addition, understanding the dynamics of a system that is simple is also usually a simple task. In other words, simple systems yield simple, predictable behaviors. For example, filling a bathtub with water requires opening the tap so that water flows into the tub at a certain rate, and restricting its outflow such that the inflow minus the outflow leads to an accumulation of water in the tub. This is a simple system. The effect of changing the rates of inflow and outflow are easily understood. Essentially there are very limited effects from limited causes that can be easily predicted. No surprising or unintended consequences result from such a simple system.

However, the behavior of a system may be simple or complex. In other words, simple systems yield simple, predictable behaviors, but complex systems often yield complex and sometimes unpredictable behavior. Generally, the complexity of behavior that a system exhibits is directly related to the complexity of the architecture of the system itself. Systems with multiple components related through complicated links present behaviors that are difficult to fully predict and understand without a mapping of the essential components and their links. In addition, it is often not possible to understand the behavior of complex systems without an actual model of the relevant portions of the system.

This is why a productive approach to the understanding of systems is based not on an aspiration to model reality but on the strategic articulation of a question that can serve to elicit a process of describing the important, relevant components of the system that may aid in answering that particular question. System dynamics is not a panacea for solving complex problems. It is also not a method that attempts to model reality in any comprehensive way. Any attempt to completely model the whole of the reality of a human situation is doomed to failure. The focus of a system dynamics approach is a question—a concern. The primary intention of a system dynamics approach is the development of an overall architecture of components that can be used not to model reality, but to model a question. In so doing, the developer may need to describe narrow or broad swaths of a situation, depending on the extent and breadth of the relevant elements. That is, if an element is part of the reality of a situation but clearly plays no role with respect to the question at hand, then there is no reason to include it in the modeling effort.

Therefore, system dynamics has been developed to assist in a methodical process of identifying the components of a system and describing, with as much specificity as possible, the nature of the links between components. Often, the goal is to derive a quantitative or mathematical relation that describes the interdependence between components of a system. Sometimes that is impossible to do, in which case a qualitative description is sought.

Using the methods of system dynamics to probe the real world can be particularly fruitful because of our inability to imagine all of the ways in which a system may behave (Forrester 1969, 1971; Sterman 2000; Barlas 2002). This seemingly uncertain behavior under a vast range of conditions lends powerful utility to the system dynamics approach matched by no other methodology. Human systems, societal structures, are exceedingly complex and difficult to predict. Today, many disciplines acknowledge that the obstacles toward real change—be it toward a sustainable society, alternative energy system, or explanation of environmental links to disease—relies on our ability to understand the dynamics of socioeconomic structures. Having greater understanding of anthropogenic constructs can be one approach toward addressing the fundamental problems of society (Barlas 2002).

To better appreciate the value of this approach, an introduction to the system dynamics perspective follows.

System Dynamics

As described above, the focus of much of systems theory are complex systems whose behaviors are difficult to fully understand. However, even this clear focus brings with it the challenge of determining the complexity of a system from its behavior.

Érdi poses the scenario of a stone as an example to illustrate the delineation of a system boundary (Érdi 2008, pp. 5–6). The following adapts and extends this useful example. As an inanimate object, it is reasonable to consider that the stone is a simple closed system. As opposed to a shoot of bamboo or a cricket, the stone does not interact with its environment. There is no material exchange—no perceptible flows. Therefore, in these terms it can be considered a closed system.

However, if thermal energy is included, the stone does interact with its environment. It will absorb and discharge heat from its environment depending on the thermal gradient between it and the space in which it finds itself. So, when heat energy is included, the stone needs to be considered an as part of an open system.

Similarly, even as an open system it is clear that this is likely a simple system. Heat flow into and out of the molecular matrix of the mineral is directly dependent on a limited number of material properties. The flow of this energy can be related to a limited set of causes, and the rate and direction of the flow can be predicted quite well. But the thermal behavior of the stone can be complicated by any number of factors not initially listed in the scenario. If the stone is composed of a nonhomogeneous matrix of minerals with diverse thermal diffusivities, the actual heat flow may take a complicated path. In addition, if the surrounding environment includes air flow as wind and radiant heat from the sun, a precise description of the thermal behavior of the stone may be rather more complicated than initially thought.

In other words, what may seem like a simple system may not be upon further examination. This again highlights the importance of using a focused, clear question as the catalyst and guide to constructing a useful architecture of the system for the purpose of understanding a particular aspect of its behavior.

This can be summarized in the following way. Simple systems are composed of single cause-and-effect relationships in which proportional change in cause results in proportional change in effect in a predictable manner. Closed systems exhibit all relevant behavior from causes and effects within the system boundary. Complex systems are composed of complex cause-and-effect relationships that include feedback loops and nonlinear causality in which the proportional changes between causes and effect may seem decoupled resulting in emergent and unpredictable behavior. Open systems interact with their environment; that is, they are affected by elements outside of their immediate system boundaries (Érdi 2008).

It is now obvious why a system dynamics approach may lend insight into the workings of urban systems. In fact, system dynamics has been applied to the urban context from the very beginnings of the establishment of the theory and methods of the field. Urban systems are dependent on multiple, often diverse variables that are both related and unrelated to each other.

In addition, emergent behavior is exceedingly difficult to imagine with urban systems because of their complexity and the prevalence of time lags from several weeks to many decades and even centuries. Take, for example, the unintended consequences of constraining urban water by directing and channeling natural streams and rivers over the course of decades. Only now, with widespread local urban water scarcities and the uncertainties to be brought about by global climate change, do we understand the depth and breadth of the unintended consequences from such infrastructure decisions.

The use of system dynamics in an urban context attempts to reveal emergent behavior in this context and indicate the possibility of important unintended consequences related to policies meant to promote all manner of urban sustainability priorities (see figure 8.4).

System Dynamics of Alternative Urban Scenarios: An Approach

In using the framework of system dynamics to understand the city and the prospects for a sustainable future it is clear that there are many important questions to ask. These questions, situated within the complex system of the urban environment, may be addressed through a system dynamics exploration.

For example, there are several large-scale questions. What is needed to achieve urban sustainability? What is the relationship between urbanization and global climate change? What kind of urban sociometabolic regime can be engineered to stabilize and reduce the ecological footprint of contemporary cities?

These questions are difficult to answer partly because of the sheer enormity of the system boundary, both in space and time, that is necessary to set up the question

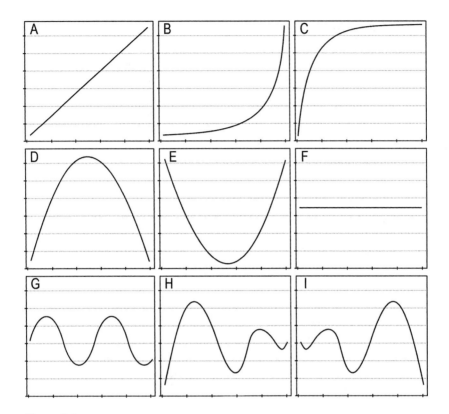

Figure 8.4
Typical system behaviors: linear increase (A), exponential increase (B), hyberbolic growth (C), growth and decline (D), decline and growth (E), no growth (F), periodic and regular oscillation (G), convergence (H), divergence (I).

itself. The number of elements within the system—approaching the entire city itself—and the span of time necessary for the study—the next several decades of global climate change, for example—make for unwieldy system architecture.

Therefore, smaller questions, focused on particular scenarios, may allow for a more productive study. These alternative urban scenarios are useful in outlining the elements that interact together within a fairly limited sector or infrastructure construct. These urban vignettes can then be described, following the priorities inherent in a particular question, and then models can be used to gain a better understanding of the governing dynamics.

9

Integrated Approaches to Sustainable Urban Metabolism

As populations in urban areas increase, urban planners and local governments are more focused to reduce the pressure on natural resource consumption by making use of more informed decision-making processes. This requires better access to more complex data representative of the sustainability challenges under a systems approach, and this can only be provided by a new framework of models integrating the different dimensions of urban issues.

Although this integrated approach is not yet available, this chapter offers a conceptual framework, including industrial ecology–oriented models, that may guide future model development under an urban systems–integrated perspective.

A Multilayer Urban Metabolism Conceptual Framework

This chapter proposes a conceptual framework to model the metabolism of urban systems. This framework is based in the assumption that the characterization of the physical nature of human economy is vital for understanding the sources and full nature of impacts of urban systems upon the natural environment, and that effective strategies toward sustainable development will rely on understanding the interaction between economic activities and their physical dimension, as represented by material and energy flows.

This framework is aimed to integrate the set of industrial ecology tools discussed in previous chapters, in order to model energy and material flows through an urban system and to correlate these flows with the economic activities that drive them. This should result from interplay between urban infrastructures, people, and communities (and their needs) and business and service providers.

The conceptual framework is offered as a set of complementary layers of models that can be used to develop high-level urban systems optimization studies and contribute to identify scenarios and subsequent policies that might arise through closing the gaps between current practice and truly material and energy-optimized cities.

From this perspective, the urban metabolism framework consists of a series of layers of models, based in different industrial ecology tools described in chapter 4, as schematically represented in figure 9.1. This framework is set up over the physical network that constitutes the urban space and its infrastructures, and the different layers share common information that builds up the understanding of the dynamics and components of the urban metabolism.

The seven layers of models that constitute this framework should be developed in order to fulfill three major goals:

1. enable and facilitate a systems perspective of the urban systems, allowing for a wide conceptual diversity of agents, multiple scales enabling multilevel analysis both in time and space, and flexibility regarding a set of interaction rules, especially between land-use including urban infrastructures, economy, and environment.

2. to be based on an analytic framework that may formalize the relationships between the different systems and processes, such as the material flows, their constituent materials, and the environmental impacts associated to its use or the drivers that determine their need.

3. to be based in statistically available data as much as possible, as this will be a key factor for a generalized utilization of these models.

Considering these major guidelines, it is suggested for the framework to be based on the physical representation of the urban space, which can be georeferenced to facilitate information storage and interaction between the different layers, whenever they require spatial resolution.

It is then proposed that the different levels of information should provide increasingly complex information, starting from urban bulk mass balances, that provide an overall balance of all the products, raw materials, and energy that enter in the urban systems, and that on the way are both stored and discarded to the surrounding environment. The second step would be to disaggregate the products in their constituent materials, as this provides a tool for planning for recycling activities or for promoting symbiosis at an urban level.

The third step should be related with adding a time dimension to the material flows analysis, which may promote the understanding of the metabolic time scales for different materials, and therefore enable an adequate planning of the infrastructure required to process them.

It is also clear that any policy intended to promote a sustainable urban metabolism is addressed to specific economic activities, and therefore the next logical level consists in providing a correlation between energy and material flows and the economic activities that are driving them.

Adding spatial resolution to the previous activities, for example by promoting them at a more granular scale, such as at the neighborhood level, can constitute the

Drivers:
- DC's unique non-state status
- Rise of mass incarceration
- history of inequality

Pressures:
- Prisons' wastewater, emissions, and other pollution issues.
- Lack of contact btwn prisoners and family
- Social inequality within DC

Impacts:
- increased rate of recidivism, loss of human capital
- Disintegration of underserved communities – exacerbate inequality
- increased environmental degradation around prisons

State:
% of prisoners relapse
% of DC neighborhoods with high incarceration rates
% of prison communities facing environmental stress

- hungry city research project: flow of people out of DC and into for profit jail in North Carolina. How much $$ was made on this flow by geo. how much was spent by DC govt. what the costs were to the families. What the police practices that contributed. Did it lead to depopulation of black neighborhoods and raising of real estate values. Did private prisons help spur gentrification in DC.

Why are prisoners sent out of DC.

Dc's origin as balancing act between slave and free states. Balancing the racial political economy of the nation. Some cities like Chicago originated because their natural location allowed them to cheaply transport goods from frontier to city.

DC existed on a different kind of frontier.

—backup: DC water

—backup: paper as key resource to govt bureaucracy. Storage of records.

— *hungryt Act. Bldg might limits.*

- Relationship of neighborhoods with high concentrations of prisoners to neighborhoods that are being re-named by real estate industry – sign of gentrification.

- Other cities – where have they traditionally housed prisoners. Where has DC traditionally housed prisoners. How close to families of prisoners.

- How many prisoners per capita did DC have historically? Today?

- Images of policing in DC

- How many cars per capita and energy use per capita for wealthy urbanites versus poor

...[illegible] ...such home. Has this coincided in any way with the real estate boom we've seen in the city and the shifting demographics of DC (away from Chocolate City)? What is the ecological impact of the building boom (more efficiency as stock is updated?), but also the new residents who arrive in these neighborhoods (more carbon intensive lifestyles of the wealthy?)?

- I think that this has negatively impacted the prisoners themselves, the prison communities where those prisoners end up, and the social ecology of DC itself.

Sources:
General:

- DC basemaps
- DC historical photos
- Maps showing demographic changes in DC
- Real estate maps of DC

Now, the flows of undesirable protesters in and out of the city in some ways masks a more insidious reality of the population flows of the city. The reality is that DC has always been marked by high degrees of social inequality. You have on one hand the elites who come there to govern, and the layers of highly educated lawyers and administrators (upper middle class) who service the government apparatus. And then you have the impoverished, uneducated descendants of slaves. No industrial base so you don't get the rise of a large educated upwardly mobile working class, that middle layer that gives a city like New York or Chicago its flavor.

So you have within the city these issues of severe inequality. So you have visible Washington (monuments, politicians, etc) – and you have what I'll call invisible Washington. You have huge destitute populations living in the shanties that front on the alleyways of the city. Slum clearances in the 1960s in Southwest. You have massive riots in the Black districts in 1968, H Street and 14th Street.

For my project I want to focus on invisible Washington, on the forces impacting this section of the city's natural resources – for people are a natural resource as well. And how the forces pushing and pulling on this natural resource in turn shape the city itself, the city's ecology, and the ecological impact of the city on the broader regions around it – the city's hinterlands.

These forces of inequality have always shaped the city – and this human landscape has shaped itself in relation to the natural landscape of the city as well. Rock Creek Park, one of the first national parks in the country, which becomes this great dividing line between rich and poor – and it remains so to this day.

And there's another important way that human geography and natural geography converge in DC in a way that's very specific to this city. And that gets back to the reality of DC as a federal city.

Kuragan – Million Dollar Blocks //   Hijacking Sustainability Adrian Parr

DC is a planned city. It was literally swampland that was chosen as the seat of government, and so the city was constructed. Unlike other places it did not grow up around the trading of a natural resource or because of its strategic location as a port or in relation to waterways etc. It never has had much of an industrial or manufacturing base. DC was not one of those US towns where fortunes were made or lost, like Chicago or New York, or Boston or San Francisco. It's never been a market center, a financial center

It's raison d'etre is politics and government.

It became a city because of the need the new US government to have an administrative center. So it's always been about administration, about bureaucracy, which is fundamentally about people. You could make an argument that paper and the storage of paper has been one of the fundamental resources and resource flows throughout the city's history.

But I want to go a different route with this project. Because government administration takes people. So I want to focus on people as the most important natural resource that defines DC as a city. And you've seen that in DC's history, where you have these population spikes every time there is an episode that requires major expansion of the federal government. Typically around wartime. Civil War – huge influx of newly freed people of African descent – former slaves. Also an expansion of government. WWI population numbers. Depression

But there are two sides to the story of people in DC. There are those there on official government business, the people who staff the huge layers of bureaucracy, etc. Then there are the people who are undesirable by the standards of those who run the government apparatus. The freed slaves. The Bonus Army. The first iteration of the March on Washington WW2.

In fact DC was chosen specifically because of population control. The threat of unruly soldiers in Philadelphia. So the federal government wanted a piece of land where everything was under its control – including the police and militia.

DC is not part of any state. Therefore it does not elect anyone to Congress or the Senate, which means that the 650,000 residents of the city have no say in the federal legislative process. Which is somewhat ironic for the capital city of a country that likes to call itself the world's greatest democracy. In fact, since 200? The license plates from DC have carried this slogan, Taxation Without Representation, which was a rallying slogan for the American War of Independence from Great Britain.

Now, many of the DC residents who come to work in the government or government related industries have recourse to voting in their home states – these are people who have roots elsewhere. But those of us who are native Washingtonians are denied and federal voting rights other than for the president. And these native Washingtonians tend to be poorer and Black.

This reality of the federal city and the unique relationship of DC to the federal government leads to other outcomes as well: specifically, I want to focus on prisons and real estate.

- A lot of attention being paid today to the issue of mass incarceration (photo of Michele Alexander's book). Prison system, huge explosion of prison population. Graph showing increase in DC prison population or other criminal justice stats.

- DC has no prisons – sends it prisoners elsewhere. What is the ecological impact of that journey? What is the ecological impact of those prisons? What is the impact of that

Prison Ecology Project: prisons "function like a small city packed into one building." "That's a lot of what prisons are: industrial warehouses that operate factories directly on the property, [including] sewage management and power plants."

Ruth Wilson Gilmore studied California's prison boom for her 2007 book Golden Gulag

Partly because of the 2001 closure of the nearby prison in Lorton, Va., which used to house D.C. inmates, prisoners are now routinely held in federal facilities hundreds of miles away from home, sometimes as far away as Florida, Texas, or California.

5,700 D.C. prisoners were housed in 33 states in facilities owned or leased by the federal government, often hundreds of miles away.

According to the Urban Institute, one in five D.C. felons is imprisoned more than 500 miles from the District.

- Shelley Broderick
- Rayyan & Bill Purdue's wife – public defenders
- Mike Stark & other crim. Justice groups in DC
  - o Institute for Policy Studies
  - o Empower DC
- Enviro justice research into prisons
- Maps showing prison locations
- Photos showing environmental impact of prisons

- Backup plan: uses of the Anacostia waterfront through time. Water in DC. Military. Power generation. Hospital. Navy Yard. Parkland. Real Estate.

My research on prisons and prisoners would look at the contracts for the transport and housing of prisoners between DC and other areas. And the real estate contracts in parts of the city with large populations caught in the prison system Imprisonment of large numbers leads to social breakdown within neighborhoods, causing indirect loss of human capital from students who do poorly in school, thus feeding back into wasteful cycle of policing and imprisonment. Causes increased fracturing of neighborhoods as they decline, and then continue to fracture with revival as whiter and wealthier residents do not integrate easily with older residents who tend to be poorer and blacker. These residents also may bring with them larger ecological footprints, as they tend to own (more) cars, consume more, etc. However real estate values also increase, and profits of private prison companies increase – what are the ecological ramifications of this? What are the ecological ramifications of sending prisoners away? Carbon footprint of that travel? Of families traveling to visit? $$ spent on trips that could have been spent otherwise? Possiblities for poisoning of inmated due to industrial processes happening at prisons? What are the impacts of the crimes committed / possibility that these are victimless crimes?

http://www.washingtoncitypaper.com/blogs/citydesk/2013/05/22/visitation-slights-how-two-policies-stack-the-deck-against-d-c-inmates/

http://www.citylab.com/crime/2015/07/how-mass-incarceration-takes-a-toll-on-the-environment/399950/

https://www.google.com/webhp?sourceid=chrome-instant&ion=1&espv=2&ie=UTF-8#q=environmental%20impacts%20of%20mass%20incarceration

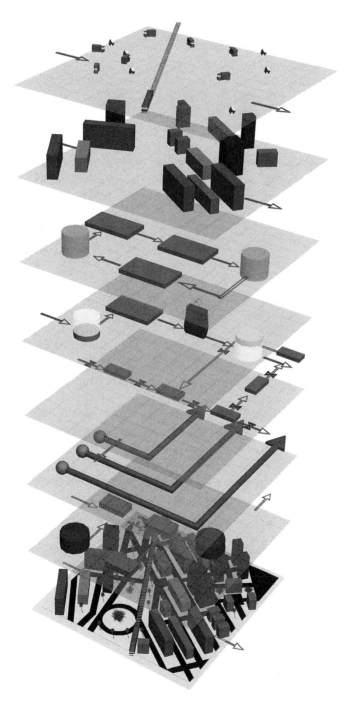

Figure 9.1
Multilayer urban metabolism conceptual framework.

next step, which could logically be accompanied with the understanding of the transport dynamics that promote most of the metabolic functions of the urban systems.

The conceptual framework suggested to model the urban metabolism includes the following layers, discussed in the sections below:

1. urban bulk mass balance
2. urban materials flow analysis
3. product dynamics
4. material intensity of economic sectors
5. environmental pressure of material consumption
6. spatial location of resource use
7. transportation dynamics.

Layer 1: Urban Bulk Mass Balance

The urban bulk mass balance layer represents the outcome of the urban metabolism, by quantifying the consumption of matter and energy its accumulation and exports, as well as the discharges of residuals in different ways, inducing environmental impacts.

An urban system is highly dependent on importing resources extracted beyond its landscape. Eventually all urban systems rely on food, fuels, and materials from elsewhere, and all cities work as marketplaces. Identically, cities usually cannot manage and dispose all the wastes they generate and do also export pollution to their hinterland. Accounting for the material flows in an urban system adopts the MFA tools described in chapter 4, with some specificity, which will be detailed in chapter 10, but overall it is assumed that there are inputs, additions to stock, outputs, and crossing flows, as represented in figure 9.2.

Urban areas are integrated in the context of a country, and therefore the distinction between imports and exports to and from foreign countries and other regions of the country have to be made. However, if at the national level the boundary between the economy and the environment is normally well defined, at the urban level, data it is not so effectively available as the transactions with the exterior represent exchanges with the surrounding municipalities and the hinterland, and this statistical data is not often recorded.

Urban bulk mass flows are based on the identification of the weight of raw materials/products that enter and leave the urban perimeter, characterizing the relationship between the urban system and its corresponding hinterland or surrounding urban systems, allowing for the assessment of its dependence in terms of supplying and disposing materials/products.

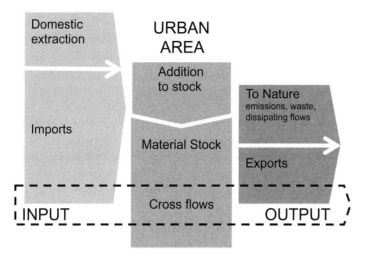

Figure 9.2
Material flows over an urban system.

In addition, while a great deal of attention has been placed on the inputs (materials and energy) and outputs (wastes) of the urban system, a focus on the potential for employing existing stocks as a buffer against systematic failure of basic urban functions might become increasingly relevant. In fact, the large material stocks of cities may, in the future, contribute to enhance the resilience of urban systems, as they may constitute a source of resources to their regular functioning and to reduce urban dependence on external resources.

In fact, the city itself represents a concentrated material stock (Obernosterer and Brunner 2001) and stocks are continuously renewed, as end-of-life products are discarded from the stock and new products are brought and become part of the stock. This renewal process defines the dynamics of the stock. In the European Union, for instance, a considerably high amount of around 60 percent of the annual direct material inputs increases the material stock of the economy (Moll, Bringezu, and Schütz 2003).

In general, there are three main possibilities of stock management: immediate recovery (e.g., recycling), long-term reuse (if a stock has a reuse potential, but no market for recycling today, a controlled landfill strategy in view of future use can be adopted), and final disposal of stock (for nonreusable materials, possibilities to achieve final storage quality must be found; Obernosterer and Brunner 2001).

Another main aspect that needs to be taken into consideration in any urban mass balance, as discussed by Rosado (2012), is that no real borders exist, and therefore, defining the amounts of products crossing the borders that are intended for endogenous consumption and not to be consumed elsewhere constitutes a major task for these models at an urban level. This problem may assume a higher dimension when the urban area serves as gateway for goods (e.g., through a big harbor, train station, or an international airport) for the country.

In addition, administrative and economically relevant cities are characterized by a considerable amount of commuters, working in the city but living nearby. If these singularities are not correctly identified, the results from these studies can overestimate the urban area material flows. This justifies that several of the urban metabolism studies have been undertaken for greater regional areas, typically corresponding to "commutersheds," hence minimizing errors associated with people moving over the boundaries on a daily basis.

This is only part of the problem related to the lack of statistical data at the municipal/regional level, and its solution requires the use of different approaches to material flows, as discussed by Rosado (2012). Different studies available in the literature focus on choosing and analyzing only the most important products/materials (Bünz Valley, Greater London, and region of York), or focus on tracing a specific substance, such as lead, copper, and phosphorus among others (Bünz Valley, City of Vienna, City of Stockholm). In doing this, several studies adopting the economy-wide MFA methodology use the statistics available at Eurostat, namely the international trade statistics and domestic extraction, and in some cases extrapolate the data for the region based on the estimations of sales, number of habitants, commuters and workers, or produced waste.

In any case, urban bulk mass balance provides an aggregated accounting with no spatial resolution and limited explanation about the drivers of consumption and how economic activities process and exchange products. In this method, it should be noted that there are major flows of materials that dominate the material balances, such as constructing materials, but some minor material flows, which might be neglected in aggregated indicators, can have large environmental impacts. As a result, this method is not directly correlated with environmental impacts, which are determined by the use of materials with different environmental effects (e.g., toxicity) and the risks associated to different technologies.

However, this method provides the first and most fundamental evidence of the flows of materials associated with urban activities, and how those might have environmental consequences, and, therefore, it can be considered the first step (the inventory phase) for a more detailed environmental analysis.

The urban bulk mass balance constitutes the first layer of the conceptual framework, and provides an aggregated vision of the material and energy balances across the urban system. Its main characteristics are represented in table 9.1.

Table 9.1
Urban bulk mass balance layer, main characteristics

	Urban bulk mass balance
Main industrial ecology tool	Bulk material flow accounting
Economic dimension	Trade (international and regional levels)
Environmental dimension	Resource use at regional level; air, water, and soil emissions at regional level
Data inputs	Products, raw materials
Data outputs	Products, wastes, and emissions
Comments	This method provides a material flow accounting of raw materials, products, wastes, and emissions, as well as the quantification of urban stocks

Layer 2: Urban Materials Flow Analysis

The MFA (materials flow analysis) represents the next logical step, in that it provides the material type discrimination to the bulk product flows calculated in the layer 1, facilitating the establishment of links between consumption and environmental impacts. The models associated to this layer are based in matrixes of composition of products and wastes that enter and leave the urban system and provide a materials flow characterization of the urban metabolism.

The material composition of products is frequently provided by industrial production surveys and manufacturers or recyclers information, and this frequently results in a significant heterogeneity in the formulation of the component materials.

In this context, MFA models, though presenting a large spectrum of approaches and parameters, have been adopting a nomenclature to designate materials that follows EUROSTATs aggregated material categories: e.g., biomass, fossil fuels, metals and non-metallic minerals, but this limits the potential to define recycling strategies, and more detailed classification would be useful.

A recent nomenclature, "MatCat," or material categorization nomenclature, together with a method to characterize it based on statistically available data, was proposed by Rosado (2012). This nomenclature represents an improvement to the Eurostat methodology for urban areas and existing urban MFA, by

• expanding the types of materials studied to a more detailed degree of material classes (twenty-eight classes);

• addressing material management issues, such as recycling potential and the economic value of waste;

• balancing the relative importance of imports and exports with the domestic extraction, in terms of their material characterization, and hence it is more adequate to urban areas;

• providing a good compromise between optimal information and the lack of information in the existing weight inventory of materials in products.

The MatCat method is based on six first level classes of materials: 1) fossil fuels, 2) metals, 3) non-metallic minerals; 4) biomass, 5) chemicals and fertilizers; and 6) others. Using these six classes of material types, further disaggregation is suggested by Rosado (2012), taking into account existing material types in the data sources as well as other specific issues. In the next paragraphs a detailed discussion on the definition of the second-level classes is made.

The second-level MatCat classes resulted from an evaluation of the combined nomenclature normally used by national statistical offices to classify commodities. The first issue relates with the multitude of different products within the same product type and the need to collect useful weight and composition data, leading to several simplifications. Furthermore, the description of the combined nomenclature allows for identifying three types of products regarding their material composition complexity: monomaterial, polymaterial, and "complex." Monomaterial products are composed almost exclusively of one type of material (e.g., tomatoes, fresh or chilled), polymaterial are products that can be described by a small set of different materials (usually two to four types—e.g., new pneumatic tires, of rubber) and "complex" are products that include several components and several materials (e.g., vacuum cleaners). The identification of materials as "complex" products leads to an additional effort to identify sources of information that could better describe them. The resulting nomenclature that constitutes MatCat is represented in table 9.2.

The MatCat nomenclature allows for the formalization of an analytic framework that quantifies the material composition of any product based on the following matrix, where $M_{n,m}$ represents the percentage of mass of material m included in product n.

$$
\mathbf{M}_{n,m} = \begin{bmatrix} m_{1,1} & m_{1,2} & \cdots & m_{1,m} \\ m_{2,1} & & \ddots & \vdots \\ \vdots & & & m_{(n-1),m} \\ m_{n,1} & m_{nm} & \cdots & m_{n,m} \end{bmatrix}
$$

with *n* representing classes of products and *m* representing types of materials. This type of tool is to be used together with the physical input-output of products of the urban system, expressed in terms of the balance of products that enter or leave the urban perimeter. As this method may offer a standard procedure, it opens the oppor-

Table 9.2
MatCat nomenclature, adapted from Rosado (2012)

Fossil fuels	Fuels
	Other fossil fuels
	Lubricants and oils and solvents
	Plastics and rubbers
Metals	Iron, steel alloying metals and ferrous metals
	Light metals
	Non-ferrous heavy metals
	Special metals
	Nuclear fuels
	Precious metals
Non-metallic minerals	Sand
	Cement
	Clay
	Stone
	Other (fibers, salt, inorganic parts of animals)
Biomass (forestry, crops, and animal products)	Agricultural biomass
	Animal biomass
	Textile biomass
	Oils and fats
	Sugars
	Wood and fuels
	Paper and board
	Non-specified biomass
Chemicals and fertilizers	Alcohols
	Chemicals and pharmaceuticals
	Fertilizers and pesticides
Others	Non-specified
	Liquids

tunity to compare results with other MFAs of cities or regions, while allowing for the calculation of different MFA indicators.

The disaggregation of the mass flows in terms of different types of materials does also facilitate the analysis of the potential to promote recycling activities that may contribute to close the materials cycles, following the industrial ecology principles.

The major improvement provided by layer 2 consists on the disaggregation of the materials that flow across and accumulate in the urban system. Layer 2 main characteristics are represented in table 9.3.

Table 9.3
Urban materials flow analysis layer, main characteristics

	Urban materials flow analysis
Main industrial ecology tool	Product composition matrixes over a bulk material flow accounting
Economic dimension	Potential for recycling activities
Environmental dimension	Material disaggregated resource use at regional level; air, water, and soil emissions at regional level
Data inputs	Product flows, product composition
Data outputs	Materials as wastes and emissions
Comments	This method provides a materials flow accounting of raw materials, products, wastes, and emissions

Layer 3: Product Dynamics

Material flow analysis is generally based on the balance of materials within a yearly base and, as a consequence, goods with a life span of more than one year do accumulate in the system, and are referred to as stocks. The stock becomes obsolete at the end of its useful life at a rate which depends on the average life span of each type of product and the distribution of products across the urban stock.

Stocks have their own characteristics and dynamic behavior. The dynamic behavior of the product stock, which is mainly determined by the behavior of the stock's inflow (purchases of new products) and stock's outflow (discarding of obsolete products) can also be described by a system dynamics model.

The outflow of the stocks depends on the mechanisms of delay and leaching. Delay represents the discarding of products and is determined by the life span of the products. Empirical data on the life span is often not available, as discussed by Elshkaki et al. (2005), and in this case either an average life span or a specific life span distribution can be assumed, such as normal, Weibull, or beta distributions. Leaching refers to the emissions of the substance from the products during the use process, which can be described as a fraction of the stock, and for different applications an emission rate can be established.

It is thus clear that if the product dynamics is to be modeled, this model should make use of a "life-span database" combined with the information of which products and goods constitute the urban stock at each moment in time.

There is some literature that discusses the importance of the life span in an industrial ecology context; see for example, Cooper (2005) or Murakami et al. (2010). The life span of commodities has also been studied for various purposes and applications, such as in marketing, where the life span of commodities has been estimated and used for forecasting the replacement market (Pennock and Jaeger 1964; Smith 1973).

Data on life span have also been used for the discussion of waste prevention and resource preservation by way of life span extension and reuse of commodities (Organization for Economic Cooperation and Development 1982; Box 1983). A number of studies over the last ten years or more have also addressed dynamic MFA and SFA (Gilbert and Feenstra 1994; Kleijn, Huele, and Van Der Voet 2000).

As discussed by Rosado (2012), the average life span of a product is affected by several variables, such as

- the product material composition, namely if it is composed of fast degrading materials (less than one year of lifetime);
- the phase of the product in the life cycle, and if they will undergo a transformation, from an intermediate state into a final product;
- the type of use, or function. For instance, there are products that become unusable after their first use (e.g., food packaging or cleaning products packaging) and others that may last decades (e.g., buildings).

This classification is important to estimate materials added to the urban stock. For example, Murakami et al. (2010) collected information spread through different fields of research and built the Lifespan database for Vehicles, Equipment, and Structures (LiVES), which catalog the life span of products and goods. More specifically, they compiled data from various sources, such as research articles, governmental and industrial association statistics, or research reports.

It can thus be concluded that a database with information on life span is a valuable tool to be combined with the stock data and a product composition database, in order to quantify the dynamics of the transformation of different products in end-of-life products that will require a processing infrastructure to transform them in valuable materials that may be re-introduced in the material urban cycle.

An additional and more sophisticated strategy that can be adopted to model the dynamic behavior of urban metabolism consists on system dynamics models which, as discussed in a previous chapter, account for flows and stocks and interpret the system behavior through time, capturing the existence of nonlinear behavior of different variables and also stocks and flows characteristics coupled with possible feedback structures. The methods of systems thinking have been used for over forty years (Forrester 1968) and can contribute to understand the causes of a dynamic problem, and this is why it becomes relevant for policy making, particularly for managing the material flows in an urban context.

In particular, the addition of time resolution to the bulk MFA analysis that is provided by this third layer of the urban metabolism framework facilitates the answer to the question: when will materials be available to be collected, as they

Table 9.4
Product dynamics layer, main characteristics

	Urban materials flow analysis
Main tools	Lifespan databases, system dynamics
Economic dimension	Infrastructures economics and recycling systems
Environmental dimension	Prediction of the amount of material flow stocks over time, as well as of end-of-life products generated annually
Data inputs	Product life span
Data outputs	Materials accumulated as stocks and end-of-life products generated annually
Comments	This method provides systems dynamic models based on product flows

reach their end of life? The information provided by this layer 3 models will not only quantify the stock is in urban areas, but will also allow quantifying when the amounts and types of materials are leaving the urban area. This is crucial to support recycling policies but also to identify the economic value of the materials in the built environment. The main characteristics of layer 3 are represented in table 9.4.

Layer 4: Material Intensity of Economic Sectors

MFA accounts are not designed to provide information on material flows at the level of economic sectors or, in particular, on inter-industry relations. Additionally they do not separate material inputs used for production processes from those directly delivered to final demand. Thus, MFA accounts and derived indicators are not able to characterize the implications of structural and technological change on resource use, as well as those associated to changes in consumption behavior and lifestyles, such as those induced by migration and urbanization. This kind of analysis is possible at the national level, making use of physical input-output tables (PIOTs), a tool which is an important extension of material flow accounts, overcoming some of the deficits identified for aggregated MFA accounts.

At an economic sector level (country scale), material inputs are primary inputs plus secondary inputs from other sectors. However, at the urban level, production sectors (industry) are scarce, exception made to the construction sector, and the main activity sectors—services—are at the end of the supply chain (final demand). Together with the households, services are *consumers* of products that are "imported" to the urban system and provided by commerce. In other words, resource consumption at the city level is due to consumption of intermediate (like the construction sector) and final consumers (services and families; see, for instance, Niza, Rosado, and Ferrão [2009]).

As opposed to the national scale, the lack of statistics at the urban level normally requires the allocation of materials per sector to be estimated. The main difficulty here is to disaggregate data in order to have higher resolution in the characterization of the activity sectors.

Rosado (2012) suggests the distribution of products per economic sector be computed making use of data provided by the international trade statistics that allocates each product into economic activities of destination, assuming that domestic products have the same destination as imports of goods.

Niza et al. (2009) provide an example of application of this methodology. It is based on the three activity sector categories suggested in the Lisbon Material Balance: a) restaurants, hotels and services, b) housing, and c) industry and construction, which is split into the Statistical Classification of Economic Activities in the European Community (NACE) and can range from one digit to five digits.

Under this framework, the economic activities distribution matrix is described in the following equation, where $AD_{n,c}$ represents the amount of product n purchased per economic activity c.

$$AD_{n,c} = \begin{bmatrix} ad_{1,1} & ad_{1,2} & \cdots & ad_{1,c} \\ ad_{2,1} & & \ddots & \vdots \\ \vdots & & & ad_{(n-1),c} \\ ad_{n,1} & ad_{n,2} & \cdots & ad_{n,c} \end{bmatrix}$$

with *n* representing classes of products and *c* representing 2-digit NACE code. This new layer, intended to quantify the material intensity of economic sectors constitutes a major policy making supporting tool as it provides data that correlates material flows with economic activities, thus leading to a more complex analysis of resource productivity in the urban space.

It is clear that this is a very relevant area that may require a significant effort in data gathering to become operational. The main characteristics of layer 4 are represented in table 9.5.

Layer 5: Environmental Pressure of Material Consumption

The extension of economic input-output tools to provide environmentally sound information is the main objective of layer 5. There have been a number of contributions in the literature attempting to use input-output (IO) techniques to calculate ecological footprints, as a major environmental pressure indicator (e.g., Bicknell et al. 1998; Lenzen and Murray 2001) or similar indicators (Proops et al. 1999; Hubacek and Giljum 2003). Other studies attempted to calculate the ecological footprint using other metrics: Bicknell et al. (1998) were the first to present an

Table 9.5
Material intensity of economic sectors layer, main characteristics

	Material intensity of economic sectors
Main industrial ecology tool	Physical input-output tables (PIOT), economic input-output (EIO) tables
Economic dimension	Characterization of the economic sectors associated material flows
Environmental dimension	Resource use at regional level
Data inputs	Economic input-output tables, materials, and energy consumption per sector
Data outputs	Materials intensity of the different economic sectors
Comments	This method provides a statistical decomposition modeling based on monetary IO tables and material flow accounting

application of input-output analysis to estimate an ecological footprint for New Zealand. The use of a single IO table serves the purpose of establishing results regarding the embodied environmental impacts in products.

The environmental extension of input-output models, namely by combining it with LCA (EIO-LCA), as discussed in chapter 4, is based on the calculation of an IO table for the economy and on the establishment of unitary environmental impacts coefficients, derived through life-cycle analysis principles.

This layer recommends the use of EIO-LCA tools in order to facilitate the management of environmental problems and economic costs, related to the increase of resources inputs and outputs. In particular, urban planning policies require answers to tangible and simple questions in order to define improvement strategies, such as: how much can we reduce in the environmental burden of urban areas, and which are the most significant products/activities and their impact? The establishment of this method is of great value to integrate economic and environmental policies at an urban level.

In EIO-LCA, input-output tables and life-cycle analysis methodologies both have a top-down approach to the problem, which tend to lead to average values and highly aggregated data, inducing several limitations to an urban model that should be considered with care, and comparisons within a time frame or between different cities might be the most relevant information that can be extracted from them. The main characteristics of layer 5 are represented in table 9.6.

Layer 6: Spatial Location of Resource Use

Space and location represent a key element for the urban metabolism approach. Incorporating information from multiple spatial scales, the analysis should reach

Table 9.6

Environmental pressure of material consumption layer, main characteristics

	Environmental pressure of material consumption
Main industrial ecology tool	EIO-LCA
Economic dimension	Trade (international and regional levels)
Environmental dimension	Resource use at regional level
Data inputs	Economic input-output tables, and environmental impact coefficients, derived through life-cycle analysis principles
Data outputs	Embodied environmental impacts per material and economic sector
Comments	This method provides embodied environmental impacts accounting based on monetary IO tables

beyond municipal boundaries in tracing the material and intangible linkages between spatially distant resources. Tracing linkages reveals the flows between different cycles and fosters awareness of the feedback mechanisms that operate across scales. This reveals the limitations of the ecosystem capacity and promotes thinking about closing linkages.

Several studies, particularly since 1973, use input-output techniques to measure the embodied energy and CO_2 of traded products and to understand the relevance of these flows in national economies. Different approaches are made, from direct estimates to the Leontieff inverse in a closed economy (Common and Salma 1992; Schaeffer and Leal de Sá 1996) and also with intercountry approaches (Lenzen 1998; Kondo, Moriguchi, and Shimizu 1998).

In general, imports to one country are due from a number of different countries and world regions with different production technologies. Each of these regions also places import demands on foreign economies. Thus, embodied production factors may continue far upstream in an international supply chain in the same way that inter-industry demands continue far upstream on the domestic level. The mathematical formulation to analyze this problem comprehensively becomes more complex. A truly multiregion input-output model is needed where inter-regional trade flows are internalized within the intermediate demand. So far, several studies have been conducted using this approach but none is still too satisfactory.

On the other hand, urban metabolism approaches should also be able to locate with the best accuracy the resources available, at the minimum spatial scale possible. Considering the allocation of products and materials, either their extraction or their consumption, within different economic activities, one can start to spatially identify their location, and their status (raw materials, used products, or waste).

Layer 6 suggests the development of a system of georeferenced socioeconomic databases, and their characterization with resource flows and stocks, to allow for the generation of a map of material flows for an urban area that can be analyzed with geographic information systems methodologies. In addition, the use of a multiregional input-output table tracing trade of products between countries will allow the generation of a global map of the interdependencies of the urban area with the world.

In order to understand the spatial distribution of materials it is crucial to have a database that characterizes the economic activities located at each chosen elementary spatial scale. This might result in a "spatial economic activities distribution matrix," which is an image of the mix of economic activities in the several spatial locations considered in the model for the urban area.

Producing this "spatial economic activities distribution matrix" may require the establishment of some rules to extrapolate the global urban data to a set of smaller areas, and these might include assuming, for example, a linear relationship between the number of workers per economic activity code in each location and the amount of products in the same location.

Knowing where, when, and how resources are used within urban boundaries will help policy makers, especially at the municipal and regional levels, to manage the needs of different zones, and can also fairly redistribute the efforts of resource allocation at an urban level. The main characteristics of layer 6 model are represented in table 9.7.

Layer 7: Transportation Dynamics

The spatial configuration of an urban system, the land-use pattern, influences and is influenced by the travel-related decisions that individuals make, which in turn reflect their activities. Modeling transport dynamics is therefore a very complex issue with huge impact and interaction on the urban metabolism, and this explains its higher-level positioning in this framework.

Table 9.7
Spatial location of resource use layer, main characteristics

	Spatial location of resource use
Main tool	Geographical information systems (GIS); spatial economic activities distribution matrix
Economic dimension	Trade (regional level)
Environmental dimension	Resource use at regional level
Data inputs	Products and location of economic activities
Data outputs	Resource consumption, waste available
Comments	This method provides spatial location of resources consumption with minimum spatial scale possible

Sivakumar (2007) provides a synthesis of transport-modeling methodologies and suggests that the development of land-use and travel-demand models can be presented in three distinct strands. One strand follows the development of travel-demand models from the early four-step models to the advanced activity-based travel-demand models. The second strand follows the development of operational integrated land use–transport models with the four-step model predominantly forming the transport component. And the third strand follows the development of advanced "next-generation" LU-T models that are disaggregate and use the activity-based approach within the transport component.

Regarding the first strand, travel-demand models, it is suggested that most travel-demand models currently in operation use a tour-based four-step modeling approach. This approach divides all individual travel into tours based at home and trips not based at home. Operational tour-based models typically consider the following home-based tour purposes—work, education, shopping, personal business, employers' business, and other. All the remaining non-home-based trips, such as a trip from the work place to lunch, or a trip from one shopping location to another, or a business trip from work, are classified under two purposes—non-home-based employer's business and non-home-based other. Within the four-step modeling framework, the frequency of these tours and trips is first predicted (known as the tour generation step). This is typically followed by mode-destination choice models or mode-destination-time period choice models in more advanced model systems (combined tour distribution and mode choice step). And finally, the network assignment procedure allocates the tours to the transport network. Tour-based models, although popular in practice, are still rather limited. They suffer from a lack of behavioral realism on several counts, many of which they share in common with the trip-based approach.

An alternative, as discussed by Sivakumar (2007), is provided by activity-based models, which solve the limitations associated with tour-based approaches. Activity-based models acknowledge the fact that the travel needs of the population are determined by their need to participate in activities spread out over time and space. Consequently, an individual's activity patterns, both in-home and out-of-home, influence the individual's travel patterns. In order to accurately quantify the travel needs of the population, it is therefore important to model the activity-travel patterns of the population (the activity-travel pattern of an individual is defined as a complete string of activities undertaken by the person over the course of a day characterized by location, time of day, and mode of travel between locations). Further, it is important to acknowledge that human beings are not islands and that they interact with each other extensively. Therefore an individual's activity-travel patterns are influenced by those of other individuals within the population, and particularly by the activity-travel patterns of other household members. There are a few operational activity-based models in use today, such as the BB System (Bowman and Ben-Akiva 2000).

On the second strand, the operational integrated land-use –transport models, we may distinguish static models, which cannot realistically capture urban spatial processes and their effects on the transport system; and dynamic models, which can be either based on general spatial equilibrium models or on agent-based microsimulation models, and both could provide significant progress to enable realistic representations of urban transportation systems.

The third strand, associated with the development of advanced "next-generation" LU-T models, according to Sivakumar (2007), may build on the strengths and experience of currently operational models, which include generally strong microeconomic formulations of land and housing/floor-space market processes and coherent frameworks for dealing with land use–transport interactions. The key tool used in these models is the microsimulation of market processes. However, of the several next-generation LU-T models currently in evolution, UrbanSim (Waddell 2002), is the only model that attempts to build integrated activity-based microsimulation models of land-use and transport.

It is thus clear that any of these options is complex but clearly interacts with the different dimensions of the urban metabolism framework and needs to be included as one of its layers.

If we turn to the facts of transportation at an urban scale, the great majority of material and energy flows within cities are carried by light transportation vehicles that represent a significant portion of all vehicle movements in urban areas: typically about 10 to 15 percent of urban vehicle trips are made for commercial purposes (Hunt and Stefan 2007).

In contrast to the regional scale, where the efficient movement of freight by heavy trucks and train, is the most common expression of commercial vehicle movements, in urban areas a much greater proportion of commercial vehicle movements are made by vehicles delivering services and goods.

However, the general state of practice in the modeling of commercial vehicle movements as part of policy analysis and planning in urban areas is rather rudimentary. There are only a few urban commercial vehicle models in use, and these often neglect important elements of urban commercial vehicle movements, including the role of service delivery, light vehicles (cars, vans, pickups, and SUVs rather than trucks) and trip-chaining (Hunt and Stefan 2007).

The most common behavior-oriented transportation models relate to "mode choice," "tour construction," and "shortest-path search"; i.e., they mainly relate to the transport demand side. In reality, however, the shippers are generally the more influential decision makers: they make decisions on shipment sizes, time windows, and other requirements that predetermine mode choice and tour construction.

Table 9.8
Transportation dynamics layer, main characteristics

	Transportation dynamics
Main tool	Agent-based models (ABM)
Economic dimension	Trade (international and regional levels)
Environmental dimension	Resource use at regional level; environmental burdens associated to transportation
Data inputs	Number of vehicles per type, km per day, type of goods transported, vehicles capacity per type, etc.
Data outputs	Transportation emissions, materials transported pathways
Comments	This method provides behavior-oriented transportation models that provide a description of transportation modes and transported goods within the urban system

Because of the strategic interactions between shippers and carriers, transport systems sometimes react in a way that cannot be predicted by studying separately both sides of the transportation market (Liedtke 2009).

Layer 7 suggests the use of different models, such as agent based models, to simulate the transportation pathways of people and goods in a way that may include information on commodities shipped, their value, weight, and mode of transportation, as well as the origin and destination of shipments for wholesale and retail.

This will provide a very valuable tool to model, with high spatial resolution, the flows of materials within the urban system, and to correlate these flows with economic activities. Layer 7's main characteristics are represented in table 9.8.

Unifying the DPSIR and the Multilayer Urban Metabolism Conceptual Frameworks

The multilayer urban metabolism conceptual framework is intended to enable a rich interaction of different models that combine the information of several layers to increase the potential of research in terms of information and decision making support.

For example, multilayer analysis may be used to propose an optimization of waste electrical and electronic equipment management by assessing the dynamics of purchase and production of end-of-life products, making use of layer 3, characterizing the added value of the industry together with environmental benefits of different waste management options by using layers 4 and 5. In addition, layers 6 and 7 could be used to design a better logistics system for the collection of end-of-life products.

This systems-based approach facilitates the design of adequate policies for this waste management at an urban level, considering both economic benefits and the minimization of environmental impacts (e.g., CO_2 emissions).

Overall, the multilayer urban metabolism conceptual framework could provide a major platform to integrate the different dimensions of an urban DPSIR framework, as discussed in chapter 1, and contribute with models that could provide adequate relationships between the DPSIR elements, and this is a major objective and contribution of this framework. Table 9.9 provides examples of indicators that could be correlated making use of the multilayer urban metabolism models, under a DPSIR framework.

Typical Data Sources

One of the major constraints to the quantification of the urban metabolism has to do with the availability of data at the urban boundaries, due to the structure of statistical records that include few data at the city level. Nevertheless, there are several important databases that can be used to extrapolate different variables in an urban context.

Data needed to account material flows of a given system include

- *inputs* domestic extraction of resources, imports of raw materials and products;
- *outputs* emissions and wastes, exports of raw materials and products.

The following data sources can be used:

- international trade: organized according to the Classification of Economic Activities and to the combined nomenclature. This includes products traded between the region and the rest of the world (excluding trade between regions of the same country).

- national freight transport: commodities transported between regions and within the studied region.

- Industrial Production Year Survey: National data about commodities produced in the country, classified according to the Classification of Economic Activities and to the combined nomenclature.

- natural resources extraction spread through different statistics, namely agriculture, fisheries, and mining.

- industrial wastes, classified according to the European Waste List and distributed per branch of the Classification of Economic Activities.

- municipal solid waste accounted by the municipalities in the urban area.

Table 9.9
Sample urban DPSIR indicators under a multilayer urban metabolism framework

	Urban bulk mass balance	Urban materials flow analysis	Product dynamics	Material intensity of economic sectors	Environmental pressure of material consumption	Spatial location of resource use	Transportation dynamics
D—Drivers	Population density, urban sprawl	Population density, urban sprawl	Technological development, market/consumer behavior	Population purchasing power, technological development, companies' activity level	Technological development, consumer behavior	Population density, urban sprawl	Population density, city economic activity
P—Pressure	Resource extraction, harvesting and fisheries, products manufacturing	Resource extraction, harvesting and fisheries, products manufacturing	Consumption of resources during products use	Energy, materials and goods consumption	Energy, materials and goods consumption	Energy, materials and goods consumption	Traffic congestion, fossil fuels consumption
S—State	Addition to stock, sealing of soil surfaces, increasing waste and emissions	Addition to stock, sealing of soil surfaces, increasing waste and emissions	Addition to stock, sealing of soil surfaces, increasing waste and emissions	Addition to stock, sealing of soil surfaces, increasing waste and emissions	Increasing waste and emissions	Addition to stock, sealing of soil surfaces, increasing waste and emissions	Air quality decrease, accessibility decrease, noise increase
I—Impact	Biodiversity decrease, critical levels of pollutants for human health	Biodiversity decrease, critical levels of pollutants for human health	Biodiversity decrease, critical levels of pollutants for human health	Biodiversity decrease, critical levels of pollutants for human health	Biodiversity decrease, critical levels of pollutants for human health	Biodiversity decrease, critical levels of pollutants for human health	Critical levels of pollutants, increase of stress-based diseases, injuries by accidents
R—Response	Urban planning, waste-management policies	Sustainable consumption policies, sustainable construction policies, waste-management policies	Quality certification, products warranty, energy labeling, sustainable consumption policies	Environmental taxes	Waste management, environmental education	Waste-management policies	Traffic management policies/regulations

Data needed to account for material composition of products include:

- identification of a material category nomenclature; and
- distribution among the categories of percentages of material weight.

Eurostat has published a guide for material flow analysis in which the nomenclature to be used is defined. Nevertheless, there is not much information available on products composition, so a thorough study must be performed to establish typical compositions of different products.

Data required to account for the life span of products is rare, although some attempts have been made; nevertheless, in this case the same thing as material composition applies. To account for economic activities several statistics are available, namely:

- number of establishments per economic activity and number of employees per economic activity; and
- population and socioeconomic information, which can be extracted from census databases.

The construction of the EIO-LCA layer is made through the use of input-output data and life-cycle analysis. Several databases are available for different environmental indicators and also for different countries. Input-output tables are constructed from national accounts statistics, and hence almost every country should have this information available.

With the gathered information and exploring the population statistics from the census dataset and business statistics, it is possible to allocate material flows to specific locations, namely to municipalities and below, which allows us to spatially describe these flows and their characteristics.

IV

Mapping and Assessing Urban Metabolism

10

Urban Metabolism in Practice: Case Studies from Developed Countries

This chapter is organized in two sections. The first analyzes some of the studies available to characterize the urban metabolism of different urban areas characteristic of Organization for Economic Cooperation and Development (OECD) countries, most of them based on material flow analysis. It is concluded that there are no harmonized methodologies that may facilitate the establishment of a comparison and benchmarking for different regions, and this constitutes the motivation for providing a detailed methodological development for the Lisbon case study, whose results are also analyzed in the context of a developed country capital.

Analysis of Developed-Country Urban Metabolism Case Studies

In the OECD countries, studies about material flow analysis have existed for some time now, mainly focused at the national level, and fewer studies focused at the regional or local level. The existing studies clearly highlight the relevance of material flows for regional and urban metabolisms (Graedel 1999). This section is dedicated to providing a brief analysis of some of the studies focused on urban areas, and the scope and main characteristics of the different approaches are highlighted.

Paul Brunner and coworkers at the Technical University of Vienna for the Bünz Valley, in Switzerland, developed one of the first urban metabolism studies (Brunner et al. 1994). In this study, water is identified as being by far the most abundant resource flow, and hence they identify the need to make a clear distinction between water and the rest of the materials. The same is true for air used in combustion processes, though at substantially lower throughput. By the time, establishing a systematic and comprehensive regional material balance was a comparatively new undertaking. The main methodological question was how to determine the important processes and fluxes in a region including thousands of anthropogenic and natural processes. The work focused on fluxes of goods and included two examples of materials such as lead and phosphorus, as hardly any methods and data existed for materials accounting at the local level.

In order to determine the flow of goods through the anthroposphere of the region, two kinds of methods were used. The process "private households" was characterized by results from existing market research studies, and the other processes were analyzed by individual surveys of the most important companies and public utilities. These techniques yield sufficient information about the flux of goods but were, in general, not applicable to the collection of data about the flow of materials, since neither private households nor many businesses knew the material composition of the goods they used. Thus, information about materials was collected from other sources, such as general tables or specific articles about the content of, say, phosphorus in goods like detergents and fertilizers.

The information about the fluxes of goods for the "average" private household and for each business was used to calculate a list of the most important processes and goods circulating in the region. Uncertainty of 20 percent is reported, as some information was missing (not all processes have been surveyed), the information does not cover the same time period for all businesses, and the data collected are not always accurate due to difficult measurements or calculations for solid waste or impossible for flue gas. At any rate, this was a pioneering study of great relevance.

Another relevant case study is that made for the city of Vienna's metabolism. For this region three different interconnected MFA investigations were carried out in order to explore the use of MFA as a decision-making tool (Hendriks et al. 2000). These were

- identification of key anthropogenic material flows and stocks within the city;
- assessment of current anthropogenic materials flowing into the environment and the effect of decisions made within the anthroposphere;
- assessment of the dependence of the city of Vienna on its hinterland, for the supply and disposal of materials (using nitrogen as an example).

This study confirms an important aspect of urban regions, namely the idea of cities as "throughflow" reactors for the most important goods (water, air, and energy sources) where the city is dependent on its hinterland for the supply, as well as the disposal, of its materials.

It does also show that urban areas are ecological "hot spots" for toxic metals, due to the dissipation of materials during the use of various products in the anthroposphere and also that the potential of the large material stocks in the built environment can represent possible resources in the future. For example, the stock of lead water pipes in Vienna's buildings was of approximately 20,000 tons, and this lead stock contains enough lead to produce approximately 1.6 million traditional car batteries.

The methodology adopted was based on the quantification of flows and stocks of goods (water, air, energy sources, producer and consumer goods, construction materials, waste and waste water), which were initially investigated in order to identify the most important substance carriers. In order to understand the urban system at a substance level, six indicator substances (carbon, nitrogen, aluminum, iron, lead and zinc) were selected for investigation and key processes and substances were selected in order to understand the urban metabolism rather than focusing on a particular environmental problem.

A different set of studies are those that follow a particular substance or good. One of these studies was focused on specific aspects of material flows, namely timber flows. This study aimed to elaborate material management models for water, biomass, construction materials, and energy, and to develop scenarios with regard to urban quality and sustainable metabolism for the urban network consisting of different towns and villages in the Swiss lowland region, Kreuzung Schweizer Mittelland (Hendriks et al. 2000). The project defined a range of criteria for sustainable development for the region, namely regional timber shortages and surpluses that should be avoided, since forecasts predict increased global shortages of timber in the twenty-first century. The mass balance and associated constraints were formulated in a mathematical model and the parameters of the model were estimated using data from local forest inventories, national censuses, and different regional and national studies regarding production, trade, consumption, and waste management.

Some of the previous studies provide brief mentions to a particular branch of material flow analysis, called substance flow analysis, and as an example, Burström, Frostell, and Mohlander (2003) developed an analysis to understand the metabolisms of nitrogen and phosphorus in the city of Stockholm study. The metabolisms of nitrogen and phosphorus were analyzed in order to present a comprehensive picture of these substances flow in the city of Stockholm for one year (1995) and provide information for the local management on related environmental issues, namely which are the main sources and underlying causes of emissions and possible sources for future emissions, together with the identification of the main sinks for material flows, and how much is accumulated in the municipality.

In different ways, these studies focused on flows and stocks to provide a tool for municipalities to properly manage material flows. We can find in these studies typical findings for urban systems: the dependence on imports of substances and the potential of the urban stocks, both for providing resources for the future and for constituting sources of toxic emissions with significant environmental consequences. Trying to improve their use and to anticipate their potential environmental problems, these methods provide estimates to account for all the stocks and flows, based on several different sources—from official statistics and scientific reports, similar studies performed elsewhere, and to a great extent by personal communication with different companies, business organizations, experts, and employees at local, regional, and national authorities.

Finally, three recent studies attempted to develop an effort to apply an MFA standardized method, first the city of Hamburg case, where the Eurostat methodology was applied, the greater London study, where there is a combination of mass balance with the ecological footprint framework, and the York study where the input-output tables framework is used.

The Hamburg study (Hammer et al. 2003) was part of the NEDS project (Sustainable development between throughput and symbolism). The main objective of NEDS consisted on the integration of ecological economics, environmental accounting (in particular MFA) discourse analysis and constructivist aspects approaches, in order to explicitly take into account the complex relationship between science and policy.

Again, the authors concluded that material flows of the city of Hamburg are almost overall dominated by its trade flows, and hence direct extraction of materials does not play an important role. They provide a discussion about the material consumption against GDP is made, trying to link resources with economic growth, which lead to the concept of dematerialization.

The Hamburg MFA followed, as far as possible, the Eurostat (2001) method. Unused domestic extraction and indirect flows associated with imports and exports have not been calculated as material input, as data was highly aggregated not allowing a useful estimation of indirect flows.

The greater London resource flows were examined (BFF, Best Foot Forward 2002) following a model proposed by Forum for the Future (Linstead and Ekins 2001). The study covers a wider range of materials, products, waste, and energy beyond the structure set out in the Linstead and Ekins (2001) model, which presents examples of a limited set of material and product flows.

The main objectives of the project were to measure resource flows and the ecological footprint for greater London, establishing scenarios to illustrate the environmental implications of a range of environmental conservation options and waste minimization and management strategies, and to assess data inputs by quality and identification data deficiencies. Part of this project aimed to identify sources of useful data and potential ways of improving data availability in the future.

According to the authors, it was not always possible to track the diverse range of material and product flows through London. This was mainly due to the difficulty in obtaining consistent data for all four data points. Efforts were concentrated on identifying "big hitter" flows, such as construction materials, which were then further investigated. Materials and products referred to in the European Commission's priority waste streams were also investigated in detail.

Finally, in the York study, Barrett et al. (2002), a MFA was made and again there were some difficulties in measuring the total amount of materials, namely the total consumption of all electrical and (nonfood) consumable items in York. An important aspect of the York study was the ability to account for "hidden flows" of materials that do not enter the economy (e.g., the removal of overburden during mining or waste trimmings from forestry). The results consisted of a set of detailed input-output tables showing the flow of materials that entered and left the city during the year 2000.

These selected works applied to different case studies show that, in principle, MFA can be applied to any well-defined economic system, be it a region, a state, or a community of states. Metabolism data have been established for only a few cities worldwide, and there are interpretation issues due to lack of common conventions.

Limits to the current applicability of this kind of study to smaller scales are associated with the availability of appropriate statistical data. The lack of available statistical data at the municipal/regional level implies different approaches to material flows at this spatial resolution. Either studies focus on choosing and analyzing only the most important products or materials (Bünz Valley, greater London, and region of York); focus on tracing a specific substance, such as lead, copper, or phosphorus, among others (Bünz Valley, city of Vienna, city of Stockholm); or analyze a sectoral material type, such as timber for the Kreuzung Schweizer Mittelland. This leads therefore to studies that generally don't explain the whole material flow within a region or city.

The obvious constraint in attempting to calculate a material flow analysis of a city is that no real borders exist; consequently, it is hard to define products trade. This problem may assume a higher dimension when the city serves as gateway for goods (e.g., through a big harbor, train station, or international airport) for the country or even for other countries (e.g., Hamburg). On the other hand, administrative and economic relevant cities are places with a considerable number of commuters, working in the city but living nearby (e.g., London).

Some particularities are also generally observed at this scale of analysis. When dealing with city material flows it is usually observed that domestic extraction is null or residual (e.g., Hamburg, York, or London). The same happens to local industrial production, and this means that hardly any raw materials are consumed within the city. Cities mainly consume final products. On the other hand, big cities usually evidence a strong relation in terms of materials or products trade with the surrounding region (cities, villages, farms, etc.) that constitute the origin and destination of many products to the city and from it.

A Methodological Development: The Lisbon Case Study

The main purpose of this study performed by Niza et al. (2009) in Lisbon was the establishment of a methodological framework for characterizing the urban metabolism, as discussed in chapter 9, whenever possible relying on published statistical data and based on the Eurostat (2001) methodology, to facilitate the "extrapolation" of the analysis. The work was focused on Lisbon city, excluding all of the surrounding municipalities of its metropolitan area.

One of the major constraints in quantifying the material flows at the urban scale had to do with the availability of data, since the structure of statistical records hardly includes data at that level. Data required to quantify the material flows of a given system include as inputs the domestic extraction of resources and imports of raw materials and products and, as outputs, the emissions and wastes and the exports of raw materials and products.

However, urban areas are usually regions where small amounts of material extraction take place. In fact, while at the end of the 1970s material extraction within Lisbon still constituted an important activity for the construction and development of the city, currently there is no active quarry in the municipality (Sousa Pinto 2005). The same is true for agricultural activity, exception made to some small private grounds and urban farming experiments.

As a consequence, it was assumed that all materials consumed in Lisbon were imported from outside its limits, including the surrounding municipalities. Therefore, import and export categories—goods exchanged between Lisbon, the rest of the country, and the rest of the world—assumed higher relevance in the quantification of the city's material flows.

On the other hand, a large number of commodities reaching the city only pass through it and should not be counted as part of the city's material flows, in order to avoid the overestimation of the material consumption associated with Lisbon residents' economic activities. For instance, Lisbon Harbor is an input and output gateway for a large number of commodities to and from the country. It processes more than 13 million tons of commodities per year, but only part of them will be consumed in Lisbon. The same is true for commodities carried by air and train.

Data related to the distribution of the labor structure and economic activities clearly indicate that Lisbon has little industrial activity. Industrial employment represented only 9 percent of total employment in 2000. Additionally, this sector registers a concentration of technologically intensive industries, which is an indicator of a specialization in knowledge-based economic activities (CML 2005). In fact, the amount of materials used for local industrial production was estimated at only 1.7 percent of the total amount of material inputs.

Since domestic extraction is almost nonexistent and local production is quite low, exports of materials derived from Lisbon were considered as residual or nonexistent, in terms of material flows. Material inputs to Lisbon were then associated with products for internal consumption and raw materials for local production systems. Outputs consisted of emissions and wastes, to be processed in the surrounding municipalities.

The remaining flows, which may constitute considerable amounts, are products that just cross through the city on the way to their destination.

In this context, the methodology developed allowed for the characterization of the following variables associated with Lisbon city material flows:

1. absolute consumption and final disposal of materials, per material category;
2. throughput of materials, per material category;
3. activity sectors' material consumption; and
4. waste treatment per material category and treatment type.

Each variable was calculated by making use of an algebraic framework based in different matrices:

1. the materials matrix provided the input and output flows of materials,
2. the throughput matrix provided the materials that are added to the city materials stock,
3. the waste treatment matrix distributed wastes according to three treatment categories, and
4. the activity sectors matrix is intended to distribute materials consumption through different economic sectors.

The first two matrices allowed for the quantification of Lisbon's material balance for 2004, and the other two were useful in characterizing the city's material consumption.

Materials Matrix

The consumption and disposal accountability was formulated as a materials matrix, which was designed as a product of three matrices:

- the products material composition matrix, A_{ij};
- the mass flow matrix, P_{jk}; and
- the Lisbon quota matrix, L_{kl}

$$M = A_{ij} \times P_{jk} \times L_{kl} \tag{10.1}$$

The products' material composition matrix, A_{ij}, where i are material categories and j are products, determines the composition of each product (input flows) and wastes (output flows) per material category, in mass percentage. Material categories considered were biomass, fossil fuels, metals, and nonmetallic minerals.

The mass flow matrix, P_{jk}, where j are products and k are mass amounts, is a diagonal matrix. Whenever $j = k$, P_{jk} defines the mass of products entering the city, in a given period, and the amounts of wastes leaving it. When $j \neq k$ then $P_{jk} = 0$.

Finally, the Lisbon quota matrix, L_{kl}, where k are mass percentages of product or waste that correspond to effective urban flows, and l are products, is also a diagonal matrix. When $k = l$, then L_{kl} defines the fraction of each product and waste that is consumed and produced within the city. When $k \neq l$, then $L_{kl} = 0$.

M, the materials matrix, results in the total amount of materials that flow (in or out) across the urban systems in a given year, and expressed in tons.

Products Composition Matrix

Material composition of products was characterized making use of the Annual Survey on Industrial Production, as this database presents data in a disaggregated way, allowing for an assessment of the composition per product, and the elemental composition was defined according to Eurostat's aggregated material categories: biomass, fossil fuels, metals, and nonmetallic minerals.

The most representative material categories in terms of weight were estimated for each set of products, grouped by the Classification of Economic Activities (CEA) and designated according to the combined nomenclature.

An identical composition for products of the same CEA was assumed and it quantified as the average value, avoiding a case-by-case analysis of about 3,000 different products. Whenever possible, manufacturers' information about product composition (e.g., chemical products) was used. In some cases, information provided by specific waste management entities was used.

This material composition was then used to characterize each input (products from international trade, national freight transport, and local industrial production) and output (industrial waste and municipal solid waste). Due to the aggregation level of international trade and national freight transport databases, average compositions of similar products previously estimated for industrial production were used to characterize them. For industrial waste, the composition per material category was based on the European List of Wastes. Lisbon's consumption of electricity was converted into the amount of fossil fuel used for this purpose.

Lisbon Quota Matrix

Since data sources used for calculations included different geographic scales, data was processed in order to characterize only urban flows. The methodology consisted of producing a diagonal matrix to quantify the fractions of products and wastes consumed and produced in the municipality, in percentage. It was assumed that a city's consumption is a function of the number of workers, the number of inhabitants, or their purchasing power.

Both international trade and national freight transport refer to the Lisbon region. To obtain the Lisbon municipality share, materials and products imported or transported regionally were distributed per local destination activity—wholesale, retail, and local production.

For wholesale and local industrial production, it was assumed that the number of workers was the limiting factor for consumption. For retail, it was considered that, in addition to the number of workers, another limiting factor was inhabitants' purchasing power. As a consequence, the value obtained by the estimation made through the ratio based on the number of workers was multiplied by the ratio of the purchasing power in the Lisbon municipality and the purchasing power of the Lisbon region.

For each import category, destination activities were grouped. Imports to the municipality (in percentages) were obtained from the ratio between the number of workers involved in these activities in the municipality and in the region.

Local production in the municipality was estimated by defining for each Industrial Production Annual Survey's activity a ratio between the national number of workers and the municipal number of workers.

By assuming that no products are exported, it was assumed that exports equal wastes produced in the city. For industrial wastes, it was assumed that production in 2004 was similar to 2003. For municipal solid wastes, the fraction of the Lisbon municipality's residents was calculated and served as a multiplier factor.

After calculating M, the total mass of flows that enter or exit the urban system, further detail was provided in terms of the dynamics of the flows, namely by quantifying the time extension of their presence in the urban system as stock (throughput matrix), the waste destinations (waste treatment matrix), and, finally, a correlation between material flows and the economic sectors that were responsible for their presence (activity sector matrix), as analyzed in the following sections.

Throughput Matrix
Throughput, as referred here, is the product life span within the city. Figures were estimated as follows:

• the product degradation characteristics—namely, whether it is composed of quickly degrading materials (less than one year); and

• the kind of use, or function. For instance, certain kinds of products become unusable after their first use (e.g., food packaging or cleaning products packaging), and others last decades (e.g., buildings).

This classification is important for the estimation of materials added to Lisbon city material stock. The material flows' life span was divided into four categories:

1. flows leaving the economy within one year after their input (e.g., food, packaging, oil as fuel),

2. flows leaving the economy after one year but within ten years after their input (relatively durable goods, e.g., toys, computers),

3. flows leaving the economy between eleven and thirty years after their input (durable goods, e.g., machines, cars, or home appliances), and

4. flows remaining in the economy for more than thirty years (long-duration goods, e.g., buildings or communication infrastructures).

Distribution of products according to their throughput velocity was computed as the product of two matrices:

- the materials matrix, M; and
- the products life span matrix, R_{ij}

$$T = M \times R_{ij} \tag{10.2}$$

In equation (10.2), the i dimension of R represents life span categories, while the j dimension represents the products. Literature with information about product life span is scarce, with the exception of the work of Cooper (2005) and Hsu and Kuo (2005). Distribution of products according to the four life span categories was therefore based in these references and own estimations.

Waste Treatment Matrix

The waste treatment matrix was designed as a product of three matrices:

- the waste composition matrix, V_{im}, where i are material categories and m are wastes, determines the composition of each waste per material category, in mass percentage;
- the waste flow matrix, X_{mn}, where m are wastes and n are mass amounts; and
- the treatment category matrix, T_{ms}, where m refers to wastes and s to the kind of treatment.

$$W = V_{im} \times X_{mn} \times T_{ms} \tag{10.3}$$

In T_{ms}, industrial waste and municipal solid waste data were distributed according to the following treatment categories: recycling, incineration (waste-to-energy recovery), and controlled landfill. Distribution by the mentioned categories was based on the Waste Affairs Institute databases.

Activity Sectors Matrix

Distribution of products per sector was computed as the product of two matrices:

- the materials matrix, M, and
- the products per sector matrix, S_{ih}

$$S = M \times S_{ih} \tag{10.4}$$

In matrix S, i represents products, and h represents sectors. Materials were distributed according to three activity sector categories: (1) restaurants, hotels, and services; (2) housing; and (3) industry and construction. Again, distribution of imports to Lisbon per sector (in percentage) was based on the number of workers in wholesale, retail, and specific industrial activities in Lisbon.

Closing the Balance

To complete the balance among material inputs, additions to stock and output emissions from fossil-fuels combustion, wastewater solid fraction, and construction and demolition wastes were calculated. Niza and Ferrão (2006) calculated the average per capita amounts of these parameters when developing Portugal material's balance. These factors were then multiplied by the Lisbon city population. Given that dissipative flows are largely due to fertilizer and pesticide use (Matthews et al. 2000), they were considered as nonexistent in the city when compared to the national scale.

Data Requirements

The data sources used to obtain this data were:

- International trade in the Lisbon region, 2004. Source: INE (National Statistics Office).
- National freight transport within the Lisbon region and between the Lisbon region and the other Portuguese regions (as delineated by the European Union NUTS II standard), 2004. Source: INE.
- Number of establishments per economic activity and number of employees per economic activity in Lisbon city, Lisbon region, and Portugal, 2003 and 2004. Source: Studies, Statistics and Planning General Directorate of the Labor and Social Security Ministry.
- Purchasing power in Lisbon city and Lisbon region. Source: INE.
- Industrial Production Annual Survey (IPAS), Portugal, 2003. Source: INE.
- Fuels sales in Lisbon city, 2003. Source: Geology and Energy General Directorate.
- Packaging waste production, 2004. Source: Portuguese Green Dot Society.
- Fisheries (main species unshipped) in Lisbon region, by harbor, 2003. Source: INE.
- Industrial wastes in Lisbon city, 2003. Source: INR (Waste Affairs Institute).
- Municipal solid waste, 2004. Source: VALORSUL.

Table 10.1
Lisbon global material flows, 2004 (1000t)

	Input		Output
Material category	Consumption	Local production[a]	Wastes and emissions
Biomass	2050	23	432[b]
Agriculture	1499	7	117
Forestry	540	16	300
Fishery	11	0	0
Fossil fuels	1190	72	1219[c]
Metallic minerals	434	34	14
Non-metallic minerals	7261	54	380
Construction	7168	51	335[d]
Industrial	85	3	34
Industrial and construction	8	0	11
Non-specified	289	3	105
Total	11223	187	2149

[a]Local production is presented in this table to illustrate that its value is residual in Lisbon— only 1.7 percent of the consumption.
[b]Includes wastewater solid fraction.
[c]Includes air emissions from fossil-fuels combustion.
[d]Construction and demolition wastes.

Lisbon Material Balance: Results

The results obtained making use of the methodology developed quantifies Lisbon material inputs as 11 million metric tons in 2004. This represents about 7 percent of the Portuguese material consumption, as represented in table 10.1.

In the same year, industrial and municipal wastes summed 625,000 metric tons. Additionally, construction and demolition wastes production was around 335,000 metric tons, wastewater solid fraction was 14,500 metric tons, and the amount of substances in air emissions was 1.2 million metric tons. Total outputs are then 2.149 million metric tons.

These results show that nonrenewable material resources represent almost 80 percent of the total material consumption, while nonmetallic minerals (mainly construction materials) are 64 percent of the nonrenewable fraction, and 11 percent are fossil fuels. The remaining 4 percent refers to metals. Renewable resources consumption (biomass) is only 18 percent of the total consumption and the remaining are nonspecified materials.

The amount of materials added to Lisbon's material stock totals about 9 million metric tons. Fast-consumption materials, not added to stock, represent about 21 percent (2.4 million metric tons) of the total input materials. These represent around 98 percent of the fast-consumption materials, and mainly include biomass and fossil fuels such as food products, gasoline, and electricity.

In the two- to ten-year life span products, biomass and fossil fuels also predominate. Here they mainly refer to wood and textile products (of natural and synthetic fibers). In the eleven- to thirty-year life span products, biomass prevails, now associated with furniture, together with metals, which are associated with equipment such as home appliances and cars. Finally, the over thirty-years life span products are essentially related to construction materials as it may be clear from the significant accumulation of nonmetallic minerals—construction materials are more than 95 percent of this category.

From the total wastes produced, 43 percent (about 270,000 metric tons) were recycled and 45 percent (about 280,000 metric tons) were incinerated producing energy. The remaining 12 percent (74,000 metric tons) were confined in a controlled landfill. Biomass values assume an important amount of the declared wastes— around 67 percent. If we consider construction and demolition waste estimations, the amount of waste totals 960,000 metric tons and the nonrenewable fraction of wastes, with recycling potential, is about 57 percent of the total production of wastes. The renewable fraction decreases to 43 percent.

The results obtained are consistent and show that the MFA methodology developed for urban metabolism quantification allows for a regular compilation of information to characterize an urban metabolism and its main functions. Constraints specific to the urban scale of analysis, and in particular, relative to an area that include harbors and airports, were overcome by the methodology developed in the study of Niza et al. (2009), and thus contribute to the establishment of a methodology that might be used in benchmarking studies over different urban areas.

11

The Challenge of Urban Metabolism in a Developing Context

The developing world will be host to a majority of the world's future urbanization. Challenges facing these cities will include providing urban residents with the critical resources necessary for humane urban living, while striving to address climate change and resource constraints. As a result, many of the cities in the developing world will find it difficult to participate in the global green city movement, because any significant commitment to participation will likely fall outside of their central mandate to provide basic services for their citizens. The challenges facing cities in the developing world, such as poverty, slum growth, crime, and inadequate access to education, health resources, and basic services, will likely overwhelm efforts to address carbon emissions and urban resource efficiency. Enormous investments in infrastructure and housing will dominate the priorities of primary, secondary, and tertiary cities in China, India, and Africa in particular. Therefore, the prospects for achieving real gains in resource efficiency and substantial contributions to lowering carbon emissions from ongoing and future urbanization in the developing world will be just as challenging, if not more so, than in the urbanized industrial and developed world. However, given the pace and scale of urbanization in the developing world, modest improvements in the resource efficiency of cities in developing regions may yield significant progress toward a sustainable future. This chapter will review the prospects for global sustainability from urbanization of the developing world.

Urban Sustainability in an Economically Developing Context

Globally, the number of people living in slums is nearly 1 billion. It is now believed that one out of every three urban residents live in a household that lacks access to clean water or sanitation, without secure tenure, and often crowded into structures that are dangerous and ill-suited to the surrounding climate. It is estimated that the population of persons living in slums will number 1.4 billion by 2020 (United Nations-Habitat 2008).

Increases in Chinese greenhouse gas emissions between 2000 and 2030 are predicted to equal the contribution from all other industrialized nations combined (Guan et al. 2009). Much of this will be due to growing affluence, partly driven by the wealth creation that comes from agglomeration economies of cities.

How do world development organizations address the persistence of slum areas in cities, especially as cities and nations become more aware and engaged in the effort to address resource constraints and global climate change? How is it possible to promote development while limiting the serious effects on the environment due to global climate change and the degradation of natural resources?

Just as the debate regarding appropriate actions in the face of global climate change has been mired in national confrontation and differing views between those in the developed industrialized nations and those in the less developed, the notion of a future urban sustainability can pit urban residents against one another, especially when a large percentage of the urban population is poor and underserved by city provisions. Therefore, considering the perspectives of developed versus developing urban contexts is important.

The Organization for Economic Cooperation and Development (OECD) was formed in 1961. The primary aims of the group are to foster strong economies in member nations. As the expansion to the G20 powerfully illustrates, the global economic landscape has shifted away from dominance by Europe, Japan, and the United States and toward China, Southeast Asia, and India. It is possible that the OECD designation may also go the way of the G8, but for the purposes of this book, the distinction between OECD and non-OECD is useful. Accession candidate (Chile, Estonia, Israel, Russia, Slovenia) and enhanced engagement countries (Brazil, China, India, Indonesia, and South Africa) are considered as outside of the OECD for the purposes of this book.

The countries that do not belong to the OECD are a fair representation of the future of the global economy. They also represent the future of economies based in cities, as most urban population growth will occur in these countries (figure 11.1). Thus, considering the cities of these countries separately leads to a good understanding of both the challenges and potential for future urban sustainability.

Five main challenges arise when considering the sustainability of urban centers located in developing regions. These challenges are part of the complex puzzle requiring attention by national and regional leaders, social and environmental organizations, and activists, researchers, and many others striving to bring both humanity and environmental responsibility to their fellow city residents.

First, urban economic development driven by rapid urbanization is seriously challenged to incorporate priorities and instruments to increase resource efficiency and value externalities. The pressures to support economic development and facilitate the growth of urban industry, with its tax revenues and employment offerings,

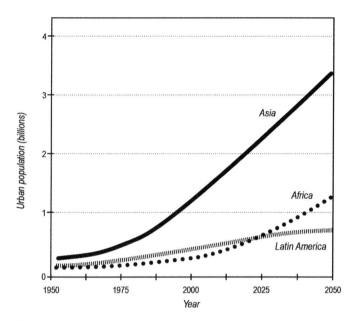

Figure 11.1
Urban population projections; Asia, Africa, and Latin America (adapted from Montgomery 2008).

often far outweigh the desire to carefully manage limited resources and honestly value the externalities of such growth. While this may seem obvious, it is not a simple challenge to address and overcome. Rapid urbanization results in enormous pressures to provide housing, employment, and services for a population substantially arising from rural to urban migration. Addressing the needs of a growing urban population is not easily dismissed or delayed, though it is clear that many cities are finding it increasingly difficult to provide even minimal levels of urban services. Providing basic services to the urban poor recently arrived and seeking employment in the city, while supporting efforts toward resource efficiency and environmental responsibility, will continue to be extraordinarily difficult.

Second, massive urbanization leads to rapid increases in the numbers of middle-class, affluent, and super-wealthy households in the city. As affluence is one of the best measures of resource intensity per capita, this increase in household wealth inevitably results in an overall increase in the resource consumption of the city. Affluence brings with it increasing demands for services of all kinds. For example, worldwide demand for air conditioning in cities is accelerating with no end in sight. China is now the leading producer of air conditioning units and, while per capita energy consumption in China lags far behind the developed world, this is likely to change at rates directly proportional to the rate of urbanization of the population. Again, the simple fact that the creation of affluence is accelerated in the innovation and business hubs of cities, and that urban affluence and access to markets drives consumption of all kinds, means that reducing the overall burden on the environment from increased consumer demand will be very difficult and run counter to trends of urban development.

Third, as outlined earlier in this book, it is now clear that the vast majority of the increase in global urban populations will occur in developing regions. Even the poorest developing areas in Africa will experience enormous urban population increases. Predicting resource scarcities and limits associated with this urbanization is a tricky business. What is not so difficult is finding examples of resource shortages and limits in urban areas today—especially in these developing regions of the world. It is therefore inevitable that growing urban populations will drive ever-critical local limits while exacerbating the various stresses on global stocks of biomass, fossil fuels, minerals, water, and other important natural capital.

Fourth, the rapid and sustained urbanization of cities in non-OECD nations of the world has led to the proliferation of informal urban settlements—the slums describe above. These informal urban areas have become a predictable and persistent pattern of urban growth and sprawl driven by massive population migration. Despite the successes of the Millennium Development Goals in reducing the proportion of the urban slum population, absolute numbers continue to increase, especially in sub-Saharan Africa. Too often policies have taken the short-term path of clearing and displacing people intent on living and working in cities, even if it means living in slum areas. Clearing residents only exacerbates the problem by catalyzing the squatting somewhere else and demonstrates the persistence of settlements with little access to services. The continued presence and growth of these urban areas has led to a general acknowledgment that formalizing these informally settled areas is one of the only ways in which to bring services—safe, reliable, and managed services—to these large populations at risk.

Fifth, the economic and social health of cities is influenced by the regional and national economies within which they reside. Cities in developing countries are subject to, and sometimes victims of, the development and trade policies of regional and national governments. Therefore, local initiatives are subject to significant pressures from centralized governments, and the international economic context within which they consider fiscal and monetary policies. A range of political, social, and cultural factors also affect the ability of cities to determine an independent path toward sustainability, even if they have the means to do so. A rare combination of elements, from civic leadership to national tolerance and support, has to come together to allow for development and implementation of urban sustainability measures.

All cities face these five challenges, though the pressures concentrated on cities in developing regions will make it more difficult to achieve consistent and strong

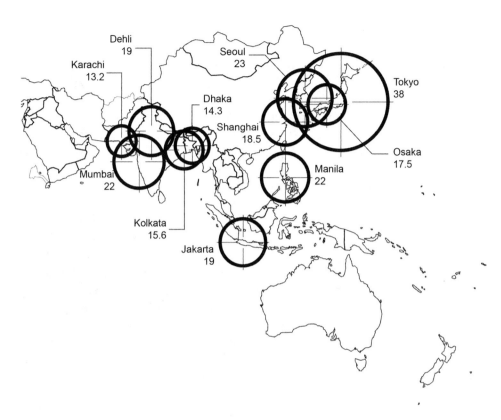

Figure 11.2
Selected cities of Asia indicating location and metropolitan populations.

movement toward urban resource efficiency. Asia is already contending with all five due to a vast urban population (see figure 11.2). How can these challenges be met in productive and practical ways? This chapter focuses on two nations and a continent; China, India, and Africa.

China

China's urban history dates back to the Zhou Dynasty of 1027 to 221 B.C., the longest-lasting dynasty. During this period, the Zhou centralized and unified a number of city-states to arrive at the founding of the first concentrated settlements. These imperial seats of power became increasingly engaged in the business of governance, as the various independent proto-feudal city-states were incorporated into the Zhou Dynasty.

Despite this ancient urban history, it is China's urbanization of the last 20 years that is truly remarkable and unprecedented in the history of the world. Between 1953 and 2003, the population of China doubled, while the urban population tripled. Six hundred and sixty cities are now home to more than half a billion of the total Chinese population of 1.3 billion. Reflecting the global urban population shift, half of the Chinese population now resides in cities. At the end of 2011, 691 million Chinese lived in cities, compared with 657 million in the countryside. In terms of overall urban population growth, the increase in the numbers of cities, expansion of urban economies and infrastructure development, the Chinese experience is like no other. China's population growth rate in 2009 was 0.51 percent, down from a high in 1966 of 2.79 percent, and population projections call for a peak population of 1.4 billion around 2030. Let's remind ourselves that the entire Chinese population in 1950 was *only* 563 million. A majority of the population now lives in the cities of the central and east regions of the country, areas that together account for two-thirds of the nation's economic activity. If the model of the United States and Europe can be applied to urbanization in China, a conservative estimate would suggest that 70 to 80 percent of the population would be accommodated in cities in the next few decades. Therefore, sometime between 2025 and 2030, the Chinese urban population will amount to somewhere between 980 million and 1.25 billion people, an increase of as many as 560 million urban residents, approximately the entire population of China in 1950.

The urbanization of China is inextricably linked to the national aspirations of the central government. Deng Xiaping, the successor to post-WWII leader Mao Zedong, began the long process of economic market-driven liberalization that has led to the vast expansion of the Chinese economy. Since 1978 the economy has expanded tenfold. Committed to continuing the expansion of gross domestic product (GDP) by at least four times, national economic goals will depend heavily on the industrial, financial, and intellectual engines of the cities. The majority of China's GDP is created in cities, and this proportion will inevitably increase (from 75 percent today to 90–95 percent in 2030; Woetzel et al. 2009). China's intensive urbanization has produced 170 cities of a million or more residents each. Several cities in these regions have been expanding at unprecedented rates. It is anticipated that sometime around 2030, over 200 Chinese cities will be home to more than 1 million residents each (CIA 2011a).

Accompanying these population increases has been a voracious land grab, by both local governments and private developers, which has extended the area of many of the largest cities well into the agricultural countryside (Liu and Diamond 2005). Early in the 1990s, urban planning initiatives to control this expansion into the countryside included ideas such as the creation of relatively small satellite cities surrounding the largest urban cores and connected with relatively compact transportation corridors. These plans were quickly overrun, however, by the sheer volume of development pressure driven by economic reform and population migration, and many Chinese cities have expanded well beyond any previous projections of area limits, by annexing townships and suburban and rural counties (Wei 2005). In addition, many previously defined counties simply acquired enough urban population and GDP to be reclassified as cities, a process of city upgrading from county level known as *xian gai shi* (Ma 2002; meaning, country becoming city).

The majority of China's existing industries and many newly founded business sectors have participated in this expansion. The construction industry in China is no exception. Construction expenditures have continually increased from US$267 billion in 2003 to a projected US$460 billion in 2008. Employment in construction has increased from 2 percent of total employment in 1990 to 5 percent in 1998. Annually, and for the foreseeable future, China builds roughly half of all new building volume in the world. From the late 1990s until 2004, China added between 1.5 billion and 2 billion m² of space annually (Lang 2004). New construction in large cities accounts for approximately 55 percent to 60 percent of this development activity. By 2020, it is estimated that 20 billion to 30 billion m² will be added to Chinese cities and suburban areas, according to Qiu Baoxing, the vice-minister of construction (Xinhua 2006). Therefore, an annual addition to building stock in China of 1.5 billion to 2 billion m² is probable until 2020.

Recently, however, the slowdown of the Chinese economy has shown that the double-digit explosion of recent years cannot carry on indefinitely. There will be significant economic structural shifts that will reflect the maturation of the consumer markets, a general rise in wages especially in manufacturing, diversification into service and high value technology development, and the ongoing development of the residential, industrial, and commercial real estate markets. Though it is widely expected that the economy will grow once again at a rate approaching or exceeding 10 percent beginning sometime in 2013, the recent financial crisis reveals weakness in exports and domestic housing markets (Organization for Economic Cooperation and Development 2011). On the one hand, this weakness readily confirms the strong links between the Chinese economy and the rest of the international, market-driven economies of the world. This is certainly a positive sign of the increasing intertwining of emerging national economies in the global economy. Despite the recent downturn, China became the world's largest exporter in 2010. On the other hand, signs of weakness signal the structural shifts listed above and others that may portend uncertainty in labor markets, intellectual capital, and industrial capacity.

Urbanization has been a primary driver of China's expanding economy partly through the creation of a wealthy, well-educated urban class that now wields significant consumer purchasing power. According to the China National Statistics

Bureau, urban disposable income has increased to more than 22,000 yuan compared with a rural cash income of 7,000 yuan, a difference of more than threefold.

China has invested heavily in the infrastructure that has supported urban development. However, the growth of the urban built environment, transportation, water and power networks, and urban industrial base are far from complete. Many thousands of kilometers of railway lines, highways, power lines, waterways, and other elements of the infrastructure are yet to be built.

Commercial and residential buildings have been the primary agents for the rapid urbanization of China. Increases in the number of commercial establishments, especially small businesses, and the exponential growth of the housing market have driven forward strong demand for sustained development and urbanization. Decoupling employment and housing has spurred a great development surge in residential construction. China devotes approximately 65 percent of its construction expenditures to improving, replacing, and adding to the housing stock (Rousseau and Chen 2001). Therefore, recent construction of the built environment in China not only has progressed at an increasing pace but has been focused on the building types of greatest material content: residential and commercial buildings (Fernández 2007).

This increase in commercial and residential buildings has begun to influence the long- term resource burden of Chinese cities. Currently China's per capita energy usage is well below most developed nations, at 50 percent of the global average. Spurred by economic development, improvements in the standard of living, increases in the volume and diversity of consumption, and large-scale migration to urban areas, per capita energy consumption will approach that of developed nations within ten to fifteen years. This will be due, in part, to the fact that in 2015 half of all residential space then existing in China will have been built since 2000 (Zhu and Lin 2004). From the year 2000 through 2030, China and India together will likely account for half of the increase in residential energy use of non-OECD countries (International Energy Outlook 2006; Fernández 2007).

China's land area of 3,705,407 mi^2 (9,596,961 km^2) is the fourth largest of all nations in the world, and just smaller than the United States. Fourteen countries share a border with China, and the total length of its coast is 9,000 miles long (14,500 km). Primary natural resources include coal and iron ore, petroleum, natural gas, mercury, tin, tungsten, antimony, manganese, molybdenum, vanadium, magnetite, aluminum, lead, zinc rare earth elements, and uranium. In addition, by way of its topography and diverse climate, China holds the world's greatest potential reserve of hydropower, with a total of one-sixth of global potential (table 11.1). Chinese hydropower production in 2007 reached 486.7 terawatt hours, a full 136.4 terawatt hours greater than the next greatest to be found in Canada.

Table 11.1
Existing hydropower resources, 2007

Province, city, district	Installed capacity (million kW)	Generating capacity per year (billion kWh)
Beijing, Tianjin, Hebei	0.43	1.2
Shanxi	0.66	1.9
Liaoning	1.07	3.7
jilin	3.38	6.9
Zhejiang	1.59	4.2
Fujian	3.34	11.8
Jiangxi	0.81	2.4
Henan	2.28	6.3
Hubei	23.47	109.0
Hubei	3.46	14.7
Guangdong	1.31	4.7
Guangxi	4.03	20.4
Sichuan	7.66	38.5
Guizhou	3.05	15.2
Yunnan	4.14	21.3
Shaanxi	1.08	4.1
Gansu	2.38	11.9
Qinghai	3.31	12.0

Source: National Bureau of Statistics of China (2007); Cai (2009).

Administration of the country is organized into twenty-three provinces (China considers Taiwan as its 23rd province), five autonomous regions, and four municipalities centrally controlled (Beijing, Shanghai, Chongqing, Tianjin) (table 11.2).

It should not be surprising to find instances in China of progressive initiatives that address the city-nature boundary as an opportunity for reconsidering urbanization and urban consumption. For example, in Chengdu, Sichuan province, a project supported by the Chengdu Urban Rivers Association (CURA) recently engaged farmers at the urban periphery to shift toward organic production to reduce pollutant run off into the river that feeds the city. CURA has even prompted the farming village of Anlong on the outskirts to set aside land for city-dwellers to rent organic farming plots. The success of this project has led to plans to build an environmental advocacy and education center called the Chengdu International Centre for Sustainable Living (You and Takelman 2011).

Finally, it is worth noting demographic trends that will affect China's urban future. The Chinese population is rapidly aging (Banister, Bloom, and Rosenberg

Table 11.2
Twenty largest Chinese cities by population of urban core

Rank	City	Population of urban core
1	Shanghai	11,740,000
2	Beijing	9,019,000
3	Hong Kong	6,780,000
4	Guangzhou	6,367,000
5	Chongqing	6,177,000
6	Tianjin	5,481,000
7	Shantou	4,936,000
8	Nanjing	4,563,000
9	Wuhan	4,516,000
10	Shenyang	4,152,000
11	Chengdu	3,958,000
12	Foshan	3,611,000
13	Harbin	3,444,000
14	Xi'an	3,269,000
15	Jinan	2,770,000
16	Qingdao	2,755,000
17	Hangzhou	2,693,000
18	Dalian	2,546,000
19	Changchun	2,542,000
20	Shijiazhuang	2,377,000

2010). The population of 1,336,718,000 (as of July 2011) has a median age of 35.5 years (34.9 yrs for males, 36.2 yrs for females), while 17.6 percent of the population is between the ages 0 and 14 yrs, 73.6 percent between 15 and 64yrs, and 8.9 percent 65 yrs and older (CIA 2011a). Chinese life expectancy is 74.68 yrs: 72.68 for males and 76.94 for females. Partly a result of China's one-child policy, the rate of the aging of the general population is one of the highest in the world.

Various attributes of cities—centralized and high-quality health care, density and access to services and markets, livability and proximity to family members—bode well for a better future for the elderly. Other aspects of urban living may pose challenges:, increases in housing and living costs generally, concentrated pollution especially unhealthy air, increased incidence of seasonal high temperatures in some city climates exacerbated by the urban heat island effect, and criminal predation on the elderly, among others. However, China retains the opportunity to guide development with an eye toward properly serving the explosion of urban elderly in the coming decades.

India

The civilization of the Indus Valley dates back to the third millennium B.C. As the Aryans from the north and northwest pushed into the land of the Dravidians around 1,500 B.C., a lasting distinction between these two tribes was created and continues today. Despite extensive mixing of these two distinct tribes and the arrival over time of Greeks, Persians, Arabs, Turks, and others, there persists a division between south and north India. Varanasi, in the north (also known as Banaras) is considered to be the oldest continuously inhabited city in India, having been settled sometime between 1,200 and 1,000 B.C.

Today, India's urbanization rivals that of China. It is one of the few countries in the world that approaches China's rate of urbanization and the size of the current and future urban population. In 2001 India's urban population amounted to 290 million, increasing to 340 million in 2008, or 30 percent of the population. By 2030, some 590 million people, 40 percent of the Indian population, will live in Indian cities, and 68 cities will be home to 1 million residents or more. This rate of urban population increase, an addition of 250 million more residents before 2030, rivals the rate at which China is adding to its urban population. As already discussed in this chapter, the future portends rapid urbanization in Asia, and India and China will be the most important countries in this future. Urbanization in India and China will result in a combined 62 percent of Asian urbanization between 2005 and 2025 (Sankhe et al. 2010).

However, today the great majority of India's population (like China's) continues to live in a rural context. Seventy percent of the population lives outside of cities, defined in India as having a population of 5,000 or more, 75 percent of male workers are employed in jobs other than agriculture, and population density is more than 400 people per square kilometer. In addition, many urban residents in India today are living in poverty, 75 percent earning less than 2 dollars per day.

India's urban future requires a focus on sustained and substantial investment in urban systems to provide for basic services and meet the needs of a growing middle class. Simultaneously, India will need to address environmental and resource issues if it hopes to avoid compounding and exacerbating those problems. Increasing infrastructure investment while managing growth in an environmental and resource-efficient way will test the resilience of municipal and regional governance (Sankhe et al. 2010).

The urbanization of India is inextricably linked to its raging economy. In 2011, India is facing challenges from high inflation resulting from years of growth. Double-digit inflation numbers have persisted for 198 months, a historically long period at such high rates. As recent as 2010, India has posted strong growth numbers, with

GDP increasing by 9.1 percent in 2009, 8.8 percent in 2010, and 7.6 percent in 2011. However, inflation has also followed suit at 7.4 percent in 2009, 10.5 percent in 2010, and 7.9 percent in 2011. Central government debt has been successfully reduced in recent years. This is an important fact owing to the need for substantial central government investment in urban India. In addition, further progress needs to be made to reform the regulatory framework for private firm involvement in the development of infrastructure. This kind of private sector investment is important for the continued development of urban infrastructure needed for supporting the growth of Indian cities. At the end of the first decade of the twenty-first century, India shows strong economic growth underpinned by substantial foreign direct investment and solid GDP growth (Organization for Economic Cooperation and Development 2011).

Cities will play a pivotal role in economic development and employment creation in the next several decades. The McKinsey Global Institute estimates that an annual projected growth rate of 7.4 percent from 2008 through 2030 will result in the creation of approximately 120 million jobs in cities. Cities will also contribute approximately 70 percent of the increase in GDP during this same period (Sankhe et al. 2010).

India's total area is 1,269,219 mi^2 (3,287,263 km^2) with a coastline measuring 4,700 miles (7,517 km) and a climate that varies from tropical monsoon in the south to temperate in the north. India possesses the fourth largest reserves of coal in the world as well as iron ore, manganese, mica, bauxite, rare earth elements, titanium ore, chromite, natural gas, petroleum, diamonds, and limestone as well as other minerals and metals. The South Asian subcontinent is flanked by the Bay of Bengal and the Arabian Sea, and its land border of 8,763 miles (14, 103 km) touches Bangladesh, Bhutan, Burma, China, Nepal, and Pakistan.

India is the world's second most populous country with 1,189,173,000 people (as of July 2011; see table 11.3). Hindus constitute 80.5 percent of the population, Muslims 13.4 percent, Christians 2.3 percent, Sikhs 1.9 percent, and others at 1.8 percent.

India's population is younger than China's; 29.7 percent are between 0 and 14 years, while 64.9 percent are between 15 and 64, and 5.5 percent 65 and older. The median age is 26.2 years, an astounding 9.1 years younger than China's median age of 35.5 years. The estimated 2011 growth rate is 1.344 percent.

India is divided into a federation of twenty-eight states and seven union territories. Each state and union territory is divided into administrative districts. All twenty-eight states and two of seven union territories are governed by elected legislatures and executive governments. The remaining five union territories are governed by the central authority through appointed administrators (CIA 2011b).

Table 11.3
Indian cities of over 1 million residents

	City	State
1	Mumbai	Maharastra
2	Calcutta	West Bengal
3	Delhi	Delhi
4	Chennai	Tamil Nadu
5	Hyderabad	Andra Pradesh
6	Bangalore	Karnataka
7	Ahmedabad	Gujarat
8	Pune	Maharashtra
9	Kanpur	Uttar Pradesh
10	Lucknow	Uttar Pradesh
11	Nagpur	Maharashtra
12	Surat	Gujarat
13	Jaipur	Rajasthan
14	Vadodara	Gujarat
15	Indore	Madya Pradesh
16	Patna	Bihar
17	Madurai	Tamil Nadu
18	Bhopal	Madya Pradesh
19	Ludhiana	Punjab
20	Coimbatore	Tamil Nadu
21	Varanasi	Uttar Pradesh
22	Vishakhapatnam	Andra Pradesh
23	Agra	Uttar Pradesh

Africa

The cities of the African continent pose particularly difficult economic and environmental challenges. With 12 percent of the urban population at risk from rising seas due to global climate change, and much of the coming increases in urban population occurring in hundreds of mid-sized cities of less than 500,000, coordinated and effective policies cannot be formulated within a context of persistent underfunding and political uncertainty (figure 11.3).

In fact, the entire notion of urban resource efficiency as a central priority of sustainable cities rings hollow when considering the facts of African life today (table 11.4). A full 40 percent of Africans continue to live under the poverty line, subsisting with under U.S. $1 per day. In addition, general access to a variety of markets and critical resources has continued to be very limited. Access to basic urban

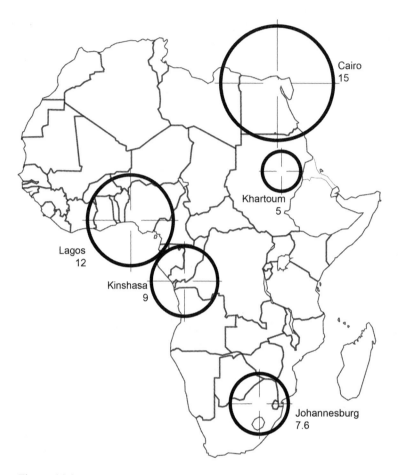

Figure 11.3
Selected cities of Africa indicating location and metropolitan populations.

services, such as water, power, and food are still a challenge for millions of urban Africans (Satterthwaite 2003). The proportion of the urban population living in slums reaches into the 80 and 90 percent range in some sub-Saharan countries. In many African cities, a small number of entities and wealthy individuals control the formal land market resulting in highly limited access for most Africans. This inability to enter into land ownership results in a chaotic informal market that promotes corruption, uncertainty, and continued economic stress for many African urban dwellers. African urban unemployment, especially for the ill-educated young, remains very high in many cities, even in the north.

Alongside these issues, a general dysfunction of national and regional governance has left many of the most rapidly urbanizing cities without the resources to attend

Table 11.4
Africa's 10 largest cities, 2005, 2010, and projected to 2025

Rank	City	Country	2005	2010	2025
1	Cairo	Egypt	10,565	11,001	13,531
2	Lagos	Nigeria	8,767	10,578	15,810
3	Kinshasa	DRC	7,106	8,754	15,041
4	Khartoum	Sudan	4,518	5,172	7,953
5	Luanda	Angola	3,533	4,772	8,077
6	Alexandria	Egypt	3,973	4,387	5,648
7	Abidjan	Côte d'Ivoire	3,564	4,125	6,321
8	Johannesburg	South Africa	3,263	3,670	4,127
9	Nairobi	Kenya	2,814	3,523	6,246
10	Cape Town	South Africa	3,091	3,405	3,824

Source: United Nations (2009, 2010).

to their growth. In several African nations, it is common that the distribution of a set of governing responsibilities to local municipal authorities is not accompanied by the necessary power or financial and professional resources required to meet the basic needs of the growing urban population. Many local authorities are charged with providing basic services, like water, but are not empowered to collect taxes to support this work. This leaves local authorities without the means to support the economic development that would result in increased municipal revenue, while tax revenue goes straight to the regional or central national government (United Nations 2010).

The urban poor experience the combined pressures of all of these issues most acutely. The welfare and compensation disparity between the urban wealthy and urban poor in Africa is the greatest in the world. In some sense this is the result of the persistence of poverty in even those African nations showing the greatest improvements in development generally; that is, wealth disparity is accentuated when some portion of the population begins the ascent up the income ladder while some portion remains at the lowest level. The southern African quadrant is the most heavily urbanized and developed region on the continent, and it can boast of some positive indicators, but also possesses one of the highest measures of wealth disparities on the continent. On the other hand, a nation with a very small wealth disparity may be evidence that very little affluence exists at all. What matters most are the consequences of pronounced welfare and compensation disparities. Are urban governments serving equally well everyone along the income ladder? Is corruption and concentration of power a norm that fundamentally affects the structure of urban land tenure and access to markets?

There have been successes in urban Africa. The African slum population has been by reduced about 24 million in the first decade of the twenty-first century. However, these reductions (a combined result of prompting through the Millennium Development Goals and various national programs) have been distributed unevenly throughout the continent. The north has shown the greatest reductions while sub-Saharan Africa lags far behind. Nigeria's urban slum population amounts to 62 percent of the total urban population, South Africa 29 percent, Egypt 17 percent (reduced from 50.2 percent in 1950), and Morocco as low as 13 percent (reduced from 37.2 percent in 1950). Despite improvements, slum improvement and elimination has continued and threatens to disrupt many aspects of the improvement of urban living and governance, and any significant moves toward urban sustainability (United Nations 2010). Urban slum populations have risen in Malawi to 70 percent of the urban population total, in Mozambique to 81 percent, and, in the Central African Republic, a staggering 96 percent! Clearly this is evidence of a complete breakdown of the urban system; a complete system crash resulting in a collapse of humane urban conditions. A sustainable resource future will have to wait for development and delivery of basic services and effective governance to take hold.

As a result, environmental health conditions in African cities continue to worsen and now affect the vast majority of the urban population. Both the conditions of poverty and development contribute to the worsening air and water quality. Under conditions of urban economic development, industrialization is distributed within the urban fabric contributing to caustic effluents to urban air and water. Under economic stasis and entrenched poverty, the use of liquid fuels and biomass for cooking and lighting contribute to poor air quality and exacerbate the incidence of asthma and other related pulmonary disorders. In either case, the lack of reliable and resilient urban services delivered via controlled and efficient infrastructure systems, along with the lack of organized industrial and household waste collection and management, create an ever-worsening urban environment.

Africa also continues to be the site of several festering and potentially metastasizing conflicts that inevitably result in disruptions to the delivery of critical resources to urban populations. From the continuing chaos of Somalia to the conflict in Sudan, internal and cross-border conflicts produce massive streams of refugees, shatter fragile plans for intra-continental infrastructure and resource management, and place governance in a permanent crisis mode. Eliminating conflict is an essential prerequisite for planning any kind of coordinated resource management on the continent.

Addressing the Challenge

Contending with the diverse challenges that belie China, India, and Africa as they continue to urbanize their populations is a priority not only for the nations directly involved but also for all other countries of the world. International trade in and of itself will make it so.

It is that urbanization in China and India is a national, and in the case of Africa, a continental phenomenon. The drivers and the effects of urbanization extend well beyond the spatial and political reach of the cities themselves. As a result, the implications of this urbanization have international and global consequences.

What is also clear in recent years is the absence of meaningful international accords on carbon emissions and resource efficiency issues. This poses a dilemma of enormous proportions. The urbanization of the developing world threatens to overwhelm efforts at global sustainability.

Despite these formidable challenges, action-oriented initiatives toward urban sustainability are under way in many developing regions. In some cases, they are supported by their national governments, in others international organizations and non-governmental groups. In some cases, there are strong and sustained grassroots movements that bring together powerful combinations of issues of social and environmental justice and environmental awareness with community mobilization and political literacy in the urban realm.

In the case of cities in developing regions, the best path forward may be unfettered urban economic development assisted by "leapfrog" technology deployment. Sustainable development that slows down growth and constrains the associated private investment and speculative entrepreneurial activities of emerging businesses may foster conditions that limit the ability of municipalities to eventually tackle the most intractable environmental, ecological, and resource challenges. It may be especially true in the developing context that a resilient city is best achieved as part of the latter stages in the development of a strong urban economy.

It is not surprising that there is a diversity of approaches to urban sustainability in the developing world. Local conditions determine the response of local and national leaders. Whether the concern is flooding in urban fringe areas, or inappropriately sited and constructed housing, the response will require practical solutions focused on short-term risk reduction with the possibility for long-term structural improvements. This is the ideal, though rarely the reality on the ground as many urban areas in developing regions suffer chronic underfunding of even the most critical aspects of municipal governance.

China, India, and Africa encapsulate the range of issues that prevail in the urban developing world. The prospects for engaging in sustainable urban solutions cannot remain elusive.

The Challenge, and Opportunities, for Urban Metabolism in a Developing Context

Much of this chapter has outlined the numerous and varied challenges facing cities in the developing world, especially under the prospect of continued massive urban population growth. Many of these challenges exist because of fundamental issues regarding the integrity of urban institutions, government, and fiscal management. These are problems that underlie the difficulties of a concerted and real move toward urban sustainability. Understandably, it is difficult to advocate for resource efficiency and environmental responsibility in a context in which basic human services and prospects for fulfilled lives is such a daunting and remote possibility for so many urban residents.

The challenge for urban metabolism work in developing contexts is thus one of relevance. Importing priorities of global concern into locations in which so much local action is required to ensure food, water, and energy security (not to mention access to markets, healthcare, and education) risks inaction and even dismissal.

Urban metabolism has a unique role to play in the urban developing context.

First, urban metabolism studies are needed that reveal the comparative resource intensity of urban places for the explicit intent of providing the most effective pathways toward resource efficiency and security to a diverse range of cities. Given the fact that the rate of growth of the population of megacities will continue to decline and huge numbers of new urban residents will be living in hundreds, if not thousands of cities of 1 million or less, regional and national governments need to provide critical development and infrastructure resources to a broad range of their cities, not just primary urban zones. A holistic understanding of the physical needs and resource intensities of efficient and humane cities can inform the distribution and implementation of this support.

Second, urban metabolism can direct attention to the most effective design and technology choices that can be made for diverse cities in developing contexts. One of the only advantages that cities in developing contexts have over their counterparts in the developed world is their position in history. They can look to the much more urbanized West and Latin America and determine the ways in which problems may be avoided now and alternative urban technologies can be adopted and developed for novel and inventive urban infrastructure. Distributed, building-integrated harvest and vehicle storage of renewable energy to support a smart, agile electricity grid is just one example of a leapfrog technology. Shared vehicles, high-yield urban farming, localized recycling, urban environmental sensing, and a variety of other technologies and design concepts can be informed by the concepts and techniques of resource accounting provided by urban metabolism.

Third, urban metabolism can inform city governments in their approach to resources in a holistic manner. Three elements are most important here. First, cities need to begin the task of collecting and organizing urban resource data for use in short-, mid-, and long-term operation and planning of cities. Second, urban metabolism can provide insight into alternative planning strategies that take greatest advantage of the "seeds" of industrial symbiosis and co-locating firms, substituting physical products and goods for urban services, and promoting the enhancement of recycling and downcycling networks toward a comprehensive closing of urban material cycles. Third, urban metabolism can assist with the multidisciplinary challenge of linking the priorities of municipal economic development with a comprehensive valuation of urban resources and resilience.

In these ways, the foundation of industrial ecology applied through the metaphor and tools of urban metabolism can help foster a constructive approach to challenges in cities of the developing world. It may be that the developing world is the best hope for a sustainable future, and it may be the cities of the developing world that hold the greatest promise.

Epilogue

Urban metabolism is a multifaceted framework in the form of a metaphor. As a metaphor, it provides a rich conceptual basis for understanding urban activities in terms of the complex flows of global resources mobilized to support them. As a methodological framework, urban metabolism provides a powerful set of tools for analyzing these flows and providing pathways toward sustainable socioeconomic structures.

Urbanization, though greatly accelerated during our fossil-fueled era, has been a hallmark of human civilization as people explored the Earth, settled in ever-larger organized groups and established the first global trade networks. The symbiosis of globalization and urbanization has defined the path of societal metabolism for much of human history. Since the earliest Neolithic settlements, cities have been founded and have grown, contracted, and died under the most diverse conditions. Their particular physical form and size, institutions and governance, and culture and social organization are as diverse as human experience. Yet they display some common attributes, including the influence that dominant resource regimes have on their morphology and infrastructure, the size and density of their resident populations, their role as centers of secular culture and religious rites, and the extent of their use and occupation of land.

Mutual benefit is the special elixir that draws people and organizations to cities. The urban world is really nothing more, and nothing less, than the accumulated and intertwined web of mutual self-interest between individuals and human organizations. Transactions of every kind abound in the urban space, and their results lead to enterprise and affluence, cultural inspiration and production, knowledge creation and innovation, and consumption as never before. While it is true that urban resource consumption is more intense now than ever before, cities have never been self-sufficient within their boundaries. They have always required a hinterland for support. As a result they are dependent on and vulnerable to changes in that hinterland, in the flux of regional and global trade and changes to local and global biogeochemical cycles.

All cities are dependent on inflows of water, food, and critical materials and are generally dominated by the kinds of energy carriers that are readily accessible within the region or through international trade. In fact, many owe their location and existence to the role they play in regional and international trade. Whether founded along waterways or overland trade routes or now acting as the hub of enormously complex digital financial networks, cities are the result of transactions at the scale of the individual and the entire world.

The ability of entire societies to purposefully alter course to avoid potential difficulties, even full-blown catastrophe, has always been limited. While there are instances when an impending disaster is tempered by the concerted actions of a government or people, there are more examples of societies failing to sense the urgency of their situation and failing to act. It seems that reconciling ourselves to a new resource and environmental reality and getting on with adaptation is a more likely path than a dramatic and transformative international effort toward structural socioeconomic and industrial change. This may be our welcome to the coming environmental disaster that some are predicting. Many would argue we are in the midst of just such a slow-motion disaster; maybe not. In this case, it is particularly unsatisfactory to state the obvious, which time will tell.

Urban metabolism, as an emerging field of studies and action, is also an indicator of the passion and motivation of a growing community that does see the challenges of our increasingly complex urban world situation. We hope that analytical understanding and the application of methodologies for assessing urban resource flows will make clear the potential for redirecting our path toward a responsible and humane future.

The authors share an optimistic regard for the benefits of our growing cities. If cities are the most complex and *human* of all of our artifacts, we can only remain optimistic that the human spirit will win out.

To a future of cities.

Notes

Chapter 2

1. "Class A" generally refers to buildings of the highest quality and best location within the commercial office real estate market.

Chapter 3

1. The Intergovernmental Panel on Climate Change has identified the built environment as demonstrating the greatest potential of all major economic sectors in delivering economically viable, global climate change mitigation strategies.

2. Fujita et al. point out that there is a real need for "modeling the sources of increasing returns to spatial concentration" (1999, 4). The approach described here is founded on this priority and therefore directs a portion of the modeling effort toward explaining changes in the urban context—physical as well as economic—based on the dynamics of agglomeration and increasing returns.

References

Adriaanse, A., S. Bringezu, A. Hammond, Y. Moriguchi, E. Rodenburg, D. Rogich, and H. Schütz. 1997. *Resource Flows: The Material Basis of Industrial Economies*. Washington: World Resources Institute.

Aggarwala, R. T., R. Desai, B. Choy, A. Fernandez, P. Kirk, A. Lazar, T. Smith, M. Watts, and A. Yan. 2011. Climate Action in Megacities: C40 Cities Baseline and Opportunities, version 1.0. London: ARUP and C40 Cities Climate Leadership Group. http://www.arup.com/Publications/Climate_Action_in_Megacities.aspx

Alberti, M. 2005. The Effects of Urban Patterns on Ecosystem Function. *International Regional Science Review* 28 (2):168–192.

Alberti, M. 2007. Ecological Signatures: The Science of Sustainable Urban Forms. *Places* 19 (3):56–60.

Alberti, M., J. M. Marzluff, E. Shulenberger, G. Bradley, C. Ryan, and C. Zumbrunnen. 2003. Integrating Humans into Ecology: Opportunities and Challenges for Studying Urban Ecosystems. *Bioscience* 53 (12):1169–1179.

Allenby, B. 1998. Context is Everything. *Journal of Industrial Ecology* 2 (2):6–8.

Allenby, B. 1999. *Industrial Ecology: Policy Framework and Implementation*. New Jersey: Prentice-Hall.

Alonso, W. 1960. A Theory of the Urban Land Market. *Papers and Proceedings Regional Science Association* 6:149–157.

Amran, M., and N. Kulatilaka. 1999. *Real Options: Managing Strategic Investment in an Uncertain World*. Boston: Harvard Business School Press.

Ausubel, J. H., and J. E. Sladovich, eds. 1989. *Technology and Environment*. Washington: National Academy Press.

Ayers, R. U., and U. E. Simonis, eds. 1994. *Industrial Metabolism*. Tokyo: United Nations University Press.

Baccini, P. and P. H. Brunner. 2012. *Metabolism of the Anthroposphere, Second Edition*. MIT Press.

Banister, J., D. E. Bloom, and L. Rosenberg. 2010. "Population Aging and Economic Growth in China." Program on the Global Demography of Aging. Working Paper no. 53. http://www.hsph.harvard.edu/pgda/WorkingPapers/2010/PGDA_WP_53.pdf

Barlas, Y. 2002. System Dynamics: Systemic Feedback Modeling for Policy Analysis. In *Knowledge for Sustainable Development—An Insight into the Encyclopedia of Life Support Systems.* vol. 1, 1131–1175. Paris : UNESCO-Eolss Publishers.

Barrett, J., H. Vallack, A. Jones, and G. Haq. 2002. A Material Flow Analysis and Ecological Footprint of York: Technical Report. Stockholm: Stockholm Environment Institute.

Batty, M. 1992. Urban Modelling in Computer-Graphic and Geographic Information Systems Environments. *Environment and Planning B* 19:663–688.

Batty, M. 2007. *Cities and Complexity: Understanding Cities with Cellular Automata, Agent-Based Models, and Fractals.* Cambridge, MA: MIT Press.

BFF (Best Foot Forward). 2002. *City Limits. A resource flow and ecological footprint analysis of Greater London.* http://www.citylimitslondon.com. Accessed December 2012.

Behrens, A., S. Giljum, J. Kovanda, and S. Niza. 2007. The Material Basis of the Global Economy: Worldwide Patterns of Natural Resource Extraction and their Implications for Sustainable Resource Policies. *Ecological Economics* 64:444–453.

Berg, P. n.d. "The Bioregional Approach for Making Sustainable Cities." Planet Drum Foundation website. http://www.planetdrum.org/bioreg_approach_cities.htm

Berg, P., and R. Dasmann. 1978. *Reinhabiting A Separate Country: A Bioregional Anthology of Northern California.* San Francisco: Planet Drum Foundation.

Bertuglia, C. S., and A. La Bella, eds. 1991. *I sistemi urbani.* Milano: Franco Angeli.

Bettencourt, L. M. A., J. Lobo, D. Helbing, C. Kühnert, and G. B. West. 2007. Growth, Innovation, Scaling, and the Pace of Life in Cities. *Proceedings of the National Academy of Sciences of the United States of America* 104 (17):7301–7306.

Bournay, E. 2008. "Urban Density and Transport-Related Energy Consumption." Chart. In UNEP/GRID-Arendal Maps and Graphics Library, http://maps.grida.no/go/graphic/urban-density-and-transport-related-energy-consumption.

Bowman, J. L., and M. E. Ben-Akiva. 2000. Activity-Based Disaggregate Travel Demand Model System with Activity Schedules. *Transportation Research* 35A:1–28.

Box, J. M. F. 1983. Extending Product Lifetime: Prospects and Opportunities. *European Journal of Marketing* 17 (4):34–49.

Breiman, L., J. H. Friedman, R. A. Olshen, C. J. Stone. 1984. *Classification and Regression Trees. The Wadsworth statistics/probability series.* Belmont, CA: Wadsworth International Group.

Bringezu, S. 1999. "Material Flow Analyses Supporting Technological Change and Integrated Resource Management." Paper presented at the Third ConAccount Meeting: Ecologizing Societal Metabolism, Amsterdam, November 21, 1998. Centre of Environmental Science, Leiden University.

Bringezu, S., and H. Schütz. 2001. "Total Material Requirement of the European Union." Technical report no. 55. Copenhagen: European Environmental Agency.

Brunner, P. H. 2002. Beyond Materials Flow Analysis. *Journal of Industrial Ecology* 6 (1):8–10.

Brunner, P. H. 2007. Reshaping Urban Metabolism. Journal of Industrial Ecology 11 (2):11–13.

Brunner, P. H., H. Daxbeck, and P. Baccini. 1994. Industrial Metabolism at the Regional and Local Level: A Case Study on a Swiss region. In *Industrial Metabolism: Restructuring for Sustainable Development*, eds. R. U. Ayres and U. E. Simonis, 163–193. Tokyo: United Nations University Press.

Burström F., B. Frostell, U. Mohlander. 2003. "Material Flow Accounting and Information for Environmental Policies in the City of Stockholm." Workshop "Quo vadis MFA? Material Flow Analysis—Where do we go? Issues, Trends and Perspectives of Research for Sustainable Resource Use," Wuppertal, Germany, October 8–10.

Cai, J. 2009. "Hydropower in China." Masters Thesis in Energy Systems, Department of Technology and Built Environment, University of Gävle, Sweden.

Câmara Municipal de Lisboa. 2005. "Diagnóstico Sócio-urbanístico da Cidade de Lisboa. Uma perspectiva censitária (2001)" [Lisbon's socio-urbanistic diagnosis. A censitarian perspective (2001)]. Colecção de Estudos Urbanos, Lisboa XXI. Câmara Municipal de Lisboa, Pelouro de Licenciamento Urbanístico e Planeamento Urbano. Lisbon: CML.

Canas, A., P. Ferrão, and P. Conceição. 2003. A New Environmental Kuznets Curve? Relationship between Direct Material Input and Income per Capita: Evidence from Industrialized Countries. *Ecological Economics* 46 (2):217–229.

CCDRLVT. 2007. "Lisboa 2020—Uma estratégia de Lisboa para a Região de Lisboa" [Lisbon 2020—A Lisbon Strategy for the Lisbon Metropolitan Region]. http://www.ccdr-lvt.pt/pt/estrategia-regional-lisboa-2020/5078.htm.

City of Cape Town. 2003. Integrated Metropolitan Environmental Policy. City of Cape Town.

City of Cape Town. 2009. City of Cape Town Environmental Agenda 2009–2014: Integrated Metropolitan Environmental Policy. http://www.capetown.gov.za/en/EnvironmentalResource Management/publications/Documents/IMEP_Env_Agenda_2009-2014.pdf

City of Cape Town. 2011a. *Smart Living Handbook: Making Sustainable Living a Reality in Cape Town Homes*. 4th ed. Cape Town: Environmental Resource Management Department, City of Cape Town.

City of Cape Town. 2011b. Supply Chain Management Policy. City of Cape Town.

Central Intelligence Agency. 2011a. *The World Factbook: East and Southeast Asia: China*. CIA online Factbook, last updated on December 20, 2011. https://www.cia.gov/library/publications/the-world-factbook/geos/ch.html

Central Intelligence Agency. 2011b. *The World Factbook: South Asia: India*. CIA online Factbook, last updated on December 20, 2011. https://www.cia.gov/library/publications/the-world-factbook/geos/ch.html

Chakravarty, S., A. Chikkatur, H. de Coninck, S. Pacala, R. Socolow, and M. Tavoni. 2009. Sharing Global CO2 Emission Reductions among One Billion High Emitters. *Proceedings of the National Academy of Sciences of the United States of America* 106 (29): 11884–11888.

Clergeau, P., J. Jokimäki, and J.-P. L. Savard. 2001. Are Urban Bird Communities Influenced by the Bird Diversity of Adjacent Landscapes? *Journal of Applied Ecology* 38:1122–1134.

Clergeau, P., J. Jokimäki, and R. Snep. 2006. Using Hierarchical Levels for Urban Ecology. *Trends in Ecology & Evolution* 21 (12):660–661.

Cleargeau, P., J.-P. L. Savard, G. Mennechez, and G. Falardeau. 1998. Bird Abundance and Diversity along an Urban-Rural Gradient: A Comparative Study between Two Cities on Different Continents. *Condor* 100:413–425.

Chertow, M. 2001. The IPAT Equation and its Variants. *Journal of Industrial Ecology* 4 (4):13–29.

Clark, W. C., and N. M. Dickson. 2003. Sustainability Science: The Emerging Research Program. *Proceedings of the National Academy of Sciences of the United States of America* 100 (14):8059–8061.

Collins, J. P., A. Kinzig, N. B. Grimm, W. F. Fagan, D. Hope, J. Wu, and E. T. Borer. 2000. A New Urban Ecology. *American Scientist* 88:416–425.

Comer, D. C. 2003. Environmental History at an Early Prehistoric Village: An Application of Cultural Site Analysis at Beidha, in Southern Jordan. *Journal of GIS in Archeology* 1 (1):105–115.

Common, M. and U. Salma. 1992. Accounting for Australian carbon dioxide emmissions. *Economic Record* 68:31–42.

Cooper, T. 2005. Slower Consumption, Reflections on Product Life Spans and the "Throwaway Society." *Journal of Industrial Ecology* 9 (1–2):51–67.

Cury, P., L. Shannon, and Y. J. Shin. 2003. "The Functioning of the Marine Ecosystems: A Fisheries Perspective." In *Responsible Fisheries in the Marine Ecosystem*, eds. M. Sinclair and G. Valdimarsson, 103–123. Rome, Italy, and Wallingford, UK: FAO and CAB International.

Daniels, P., and S. Moore. 2002a. Approaches for Quantifying the Metabolism of Physical Economies. Part I: Methodological Overview. *Journal of Industrial Ecology* 5 (4):69–93.

Daniels, P., and S. Moore. 2002b. Approaches for Quantifying the Metabolism of Physical Economies. Part II: Review of Individual Approaches. *Journal of Industrial Ecology* 6 (1):65–88.

Decker, E. H., A. J. Kerkhoff, and M. E. Moses. 2007. Global Patterns of City Size Distributions and their Fundamental Drivers. *PLoS ONE* 9:e934.

De Neufville, R., and S. Scholtes. 2011. *Flexibility in Engineering Design*. Cambridge, MA: The MIT Press.

De Neufville, R., S. Scholtes, and T. Wang. 2006. Valuing Real Options by Spreadsheet: Parking Garage Case Example. *Journal of Infrastructure Systems* 12 (3):107–111.

Dickens, C. 1877. *Our Mutual Friend*. Cambridge, MA: Riverside Press.

Dixit, A., and R. Pindyck. 1994. *Investment under Uncertainty*. Princeton: Princeton University Press.

Dolley, T. P. 2010. Minerals Yearbook: Stone, Dimension. Advance Release. Washington: Publication of the U.S. Geological Survey, U.S. Department of the Interior.

Ehrenfeld, J. 2000. Industrial Ecology: Paradigm Shift or Normal Science? *American Behavioral Scientist* 44 (2):229–244.

Ehrenfeld, J. R. 2004. Can Industrial Ecology be the "Science of Sustainability?" *Journal of Industrial Ecology* 8 (1–2):1–3.

Ehrenfeld, J. 2008. *Sustainability by Design: A Subversive Strategy for Transforming Our Consumer Culture*. New Haven: Yale University Press.

Elshkaki, A., E. Van der Voet, V. Timmermans and M. Van Holderbeke. 2005. Dynamic stock modeling: A method for the identification and estimation of future waste streams and emissions based on past production and product stock characteristics. *ENERGY* 30: 1353–1363.

Erb, K.-H., S. Gingrich, F. Kraasmann, and H. Haberl. 2008. Industrialization, Fossil Fuels, and the Transformation of Land Use. *Journal of Industrial Ecology* 12 (5/6):686–703.

Érdi, P. 2008. *Complexity Explained*. Berlin: Springer Verlag.

European Environment Agency. 1998. "Material Flow-Based Indicators in Environmental Reporting." Environmental Issues Series, no. 14. Luxembourg.

European Topic Centre on Waste and Material Flows. 2003. "Zero Study: Resource Use in European Countries. An Estimate of Materials and Waste Streams in the Community, including Imports and Exports using the Instrument of Material Flow Analysis." Copenhagen: ETC-WMF.

Eurostat. 2001. "Economy-Wide Flow Accounts and Derived Indicators. A Methodological Guide." Luxembourg: European Communities.

Fernández, J. E. 2007. Resource Consumption of New Urban Construction in China. *Journal of Industrial Ecology* 11 (2):99–115.

Fischer-Kowalski, M. 1998. Society's Metabolism: The Intellectual History of Materials Flow Analysis, Part I, 1860–1970. *Journal of Industrial Ecology* 2 (1):61–78.

Fischer-Kowalski, M., and W. Hüttler. 1999. Society's Metabolism: The Intellectual History of Materials Flow Analysis, Part II, 1970-1998. *Journal of Industrial Ecology* 2 (4): 107–136.

Flyvbjerg, B., N. Bruzelius, and W. Rothengatter. 2003. *Megaprojects and Risk: An Anatomy of Ambition*. Cambridge, UK: Cambridge University Press.

Ford, D., D. Lander, and J. Voyer. 2002. A Real Options Approach to Valuing Strategic Flexibility in Uncertain Construction Projects. *Construction Management and Economics* 20: 343–351.

Ford, R. 2008. *Who's Your City?* Philadelphia: Basic Books.

Forrester, J. W. 1968. *Principles of Systems*. Cambridge, MA: Wright Allen Press.

Forrester, J. W. 1969. *Urban Dynamics*. Portland: Productivity Press.

Freeman, H. 1987. Mental Health and Urban Policy. *Cities (London, England)* 4 (2): 106–111.

Freire, F., P. Ferrao, C. Reis, and S. Thore. 2000. "Life Cycle Activity Analysis Applied to the Portuguese Used Tire Market." SAE Technical Paper Series, no. 2000–01–1507. Warrendale, PA: Society of Automotive Engineers.

Freire, F., S. Thore, and P. Ferrão. 2001. Life Cycle Activity Analysis: Logistics and Environmental Policies for Bottled Water in Portugal. *OR-Spektrum* 23 (1):159–182.

Fujita, M., P. Krugman, and A. Venables. 1999. *The Spatial Economy*. Cambridge, MA: The MIT Press.

Gilbert, A. J., and J. F. Feenstra. 1994. A Sustainability Indicator for the Dutch Environmental Policy Theme "Diffusion": Cadmium Accumulation in Soil. *Ecological Economics* 9 (3): 253–265.

Giljum, S., and K. Hubacek. 2009. Conceptual Foundations and Applications of Physical Input-Output Tables. In *Handbook of Input-Output Economics in Industrial Ecology*, ed. S. Suh, 61–75. Springer.

Glaeser, E. 1998. Are Cities Dying? *Journal of Economic Perspectives* 12 (2):139–160.

Glaeser, E. L. 2008. "The Economic Approach to Cities." Harvard Institute of Economic Research Discussion Paper No. 2149; KSG Working Paper No. RWP08–003. Available at SSRN: http://papers.ssrn.com/sol3/papers.cfm?abstract_id=1080294.

Glaeser, E. 2011a. *Triumph of the City: How Our Greatest Invention Makes Us Richer, Smarter, Greener, Healthier, and Happier*. New York: Penguin.

Glaeser, E. L. 2011b. Cities, Productivity, and Quality of Life. *Science* 333 (6042):592–594.

Glaeser, E. 2011c. Engines of Innovation. *Scientific American* September:50–55.

Glaeser, E. L., and M. E. Kahn. 2008. "The Greenness of Cities: Carbon Dioxide Emissions and Urban Development." Harvard Institute of Economic Research Discussion Paper No. 2161. Available at SSRN: http://papers.ssrn.com/sol3/papers.cfm?abstract_id=1204716.

Graedel, T. E. 1999. Industrial Ecology and the Ecocity. *Bridge* 29 (4):4–9.

Greden, L. 2005. "Flexibility in Building Design: A Real Options Approach and Valuation Methodology to Address Risk." PhD Dissertation, Building Technology Program, Department of Architecture, MIT.

Grimm, N. B., S. H. Faeth, N. E. Golubiewski, C. L. Redman, J. Wu, X. Bai, and J. M. Briggs. 2008. Global Change and the Ecology of Cities. *Science* 319:756–760.

Guan D., G. Peters, C. Weber, and K. Hubacek. 2009. Journey to World Top Emitter—An Analysis of the Driving Forces of China's Recent Emissions Surge. *Geophysical Research Letters* 36 (4):1–5.

Gutman, P. 2007. Ecosystem Services: Foundations for a New Rural-Urban Compact. *Ecological Economics* 62:383–387.

Haberl, H., M. Fischer-Kowalski, F. Krausmann, H. Weisz, and V. Winiwarter. 2004. Progress toward Sustainability? What the Conceptual Framework of Material and Energy Flow Accounting (MEFA) Can Offer. *Land Use Policy* 21:199–213.

Hammer, M., S. Giljum, S. Bragigli, and F. Hinterberger. 2003. "Material Flow Analysis on the Regional Level: Questions, Problems, Solutions." NEDS working paper no. 2. Hamburg, Germany.

Hashimoto, S., H. Tanikawa, and Y. Moriguchi. 2007. Where Will Large Amounts of Material Accumulated in the Economy Go? A Material Flow Analysis of Construction Materials for Japan. *Waste Management (New York, N.Y.)* 27:1725–1738.

Hendriks, C., D. Müller, S. Kytzia, P. Baccini, and P. Brunner. 2000. Material Flow Analysis: A Tool to Support Environmental Policy Decision Making. Case Studies on the City of Vienna and the Swiss Lowlands. *Local Environment* 5 (3):311–328.

Hepinstall, J. A., M. Alberti, and J. M. Marzluff. 2008. Predicting Land Cover Change and Avian Community Responses in Rapidly Urbanizing Environments. *Landscape Ecology* 23:1257–1276.

Ho, S., and L. Liu. 2003. How to Evaluate and Invest in Emerging A/E/C Technologies under Uncertainty. *Journal of Construction Engineering and Management* 129 (1):16–24.

Hoffrén, J., J. Luukkanen, and J. Kaivo-oja. 2001. Decomposition Analysis of Finnish Material Flows: 1960-1996. *Journal of Industrial Ecology* 4 (4):91–112.

Horvath, A. 2004. Construction Materials and the Environment. *Annual Review of Environment and Resources* 29:181–204.

Hubacek, K. and S. Giljum 2003. Applying Physical Input-Output Analysis to Estimate Land Appropriation (Ecological Footprints) of International Trade Activities. *Ecological Economics* 44 (1):137–151.

Intergovernmental Panel on Climate Change. 2007. "Climate Change 2007: Synthesis Report. Summary for Policymakers." Geneva: World Meteorological Organization.

International Energy Outlook. 2006. *International Energy Outlook 2006*. Washington, DC: Energy Information Administration.

Jacobs, J. 1961. *The Death and Life of Great American Cities*. New York: Vintage Books.

Kaye, J. P., P. M. Groffman, N. B. Grimm, L. A. Baker, and R. V. Pouyat. 2006. A Distinct Urban Biogeochemistry? *Trends in Ecology & Evolution* 21 (4):192–199.

Kennedy, C., J. Cuddihy and J. Engel-Yan,. 2007. The Changing Metabolism of Cities. *Journal of Industrial Ecology* 11 (2):43–59.

Kitzes, J., A. Galli, S. M. Rizk, A. Reed, and M. Wackernagel. 2008. Guidebook to the National Footprint Accounts: 2008 Edition. Oakland: Global Footprint Network. Available at www.footprintnetwork.org.

Kleijn, R., R. Huele, and E. Van Der Voet. 2000. Dynamic Substance Flow Analysis: The Delaying Mechanism of Stocks, with the Case of PVC in Sweden. *Ecological Economics* 32 (2):241–254.

Koenig, H. E., and J. E. Cantlon. 2000. Quantitative Industrial Ecology and Ecological Economics. *Journal of Industrial Ecology* 3 (2–3):63–83.

Kondo, Y., Y. Moriguchi. and H. Shimizu 1998. CO2 Emissions in Japan: Influences of Imports and Exports *Applied Energy* 59 (2–3):163–174.

Krausmann, F., M. Fischer-Kowalski, H. Schandl, and N. Eisenmenger. 2008. The Global Sociometabolic Transition: Past and Present Metabolic Profiles and Their Future Trajectories. *Journal of Industrial Ecology* 12 (5/6):637–656.

Krausmann, F., S. Gingrich, N. Eisenmenger, K. H. Erb, H. Haberl, and M. Fischer-Kowalski. 2009. Growth in Global Materials Use, GDP and Population During the 20th Century. *Ecological Economics* 68 (10):2696–2705.

Lang, S. 2004. Progress in Energy-Efficiency Standards for Residential Buildings in China. *Energy and Building* 36 (12):1191–1196.

Lenzen, M. 1998. Primary energy and greenhouse gases embodied in Australian final consumption: an input-output analysis. *Energy Policy* 26:495–506.

Leontief, W., ed. 1986. *Input-Output Economics*. 2nd ed. New York: Oxford University Press.

Leviakangas, P., and J. Lahesmaa. 2002. Profitability Evaluation of Intelligent Transport System Investments. *Journal of Transportation Engineering* 128 (3):276–286.

Linstead, C., and P. Ekins. 2001. "Mass Balance, UK Mapping UK Resource and Material Flows." London: Royal Society for Nature Conservation.

Liu, J., and J. Diamond. 2005. China's Environment in a Globalizing World. *Nature* 435 (30):1179–1186.

Loh, W. Y. 2009. Improving the Precision of Classification Trees. *Annals of Applied Statistics* 3 (4):1710–1737.

Ma, L. J. C. 2002. Urban Transformation in China, 1949–2000. *Environment and Planning* 34 (9):1545–1569.

Maddison, A. 2007. Population Growth over the last 500 Years. http://visualizingeconomics.com/blog/2007/12/09/comparing-population-growth-china-india-africa-latin-america-western-europe-united-states. Accessed on December 12, 2012.

Mäler, K.-G. 2000. Development, Ecological Resources and their Management: A Study of Complex Dynamic Systems. *European Economic Review* 44:645–665.

Matthews, E., C. Amann, S. Bringezu, M. Ficher-Kowalski, W. Hüttler, R. Kleijn, Y. Moriguchi, et al. 2000. *The Weight of Nations: Material Outflows from Industrial Economies*. Washington: World Resources Institute.

McNeill, J. R. 2000. *Something New Under the Sun: An Environmental History of the Twentieth-Century World*. New York: Norton.

Miller, R., and P. Blair. 1985. *Input-Output Analysis: Foundation and Extension*. Englewood Cliffs, NJ: Prentice-Hall.

Mills, E. S. 1967. An Aggregative Model of Resource Allocation in a Metropolitan Area. *American Economic Review* 57: 197–210.

Ministry for the Environment, New Zealand. 2008. Characteristics of Sustainable and Successful Urban Areas. http://www.mfe.govt.nz/issues/urban/sustainable-development/characteristics-areas.html#ftn13r.

Moll, S., S. Bringezu, and H. Schütz. 2003. *Zero Study: Resource Use in European Countries. An Estimate of Materials and Waste Streams in the Community, Including Imports and Exports Using the Instrument of Material Flow Analysis*. Copenhagen: European Topic Centre on Waste and Material Flows.

Montgomery, M. R. 2008. The Urban Transformation of the Developing World. *Science* 319:761–764.

Mulligan, G. F. 1984. Agglomeration and Central Place Theory: A Review of the Literature. *International Regional Science Review* 9:1–42.

Murakami, S., M. Oguchi, T. Tasaki, I. Daigo, and S. Hashimoto. 2010. Lifespan of Commodities, Part I. The Creation of a Database and Its Review. *Journal of Industrial Ecology* 14 (4):598–612.

Muth, R. 1968. *Cities and Housing*. Chicago: University of Chicago Press.

National Bureau of Statistics of China. 2007. *China Statistical Yearbook*. Beijing: NBSC.

Newman, P. W. G. 1999. Sustainability and Cities: Extending the Metabolism Model. *Landscape and Urban Planning* 44 (4):219–226.

Newman, P. and I. Jennings. 2008. *Cities As Sustainable Ecosystems: Principles and Practices*. Island Press.

Newman, P., and J. R. Kenworthy. 1988. The Transport Energy Trade-Off: Fuel-Efficient Traffic versus Fuel-Efficient Cities. *Transportation Research* 22A (3):163–174.

Niza, S., and P. Ferrão. 2005. "Material Flow Accounting Tools and its Contribution for Policy Making." Paper presented at 6th International Conference of the European Society for Ecological Economics, Lisbon, Portugal, June 14–17.

Niza, S., and P. Ferrão. 2006. A Transitional Economy's Metabolism: The Case of Portugal. *Resources, Conservation and Recycling* 46:265–280.

Niza, S., L. Rosado, and P. Ferrão. 2009. Urban Metabolism: Methodological Advances in Urban Material Flow Accounting Based on the Lisbon Case Study. *Journal of Industrial Ecology* 13 (3):384–405.

Obernosterer, R., and P. H. Brunner. 2001. Urban Metal Management: The Example of Lead. Journal of Water, Air, & Soil Pollution: *Focus (San Francisco, Calif.)* 1 (3-4):241–253.

Odum, E. P. 1953. Fundamentals of Ecology. Independence, KY: Brooks Cole.

Odum, E. P. 1969. The Strategy of Ecosystem Development. *Science* 126:262–270.

Odum, H. T. 1971. *Environment, Power and Society.* New Jersey: Wiley-Interscience.

Odum, H. T., and E. C. Odum. 2006. The Prosperous Way Down. *Energy* 31:21–32.

Olalla-Tárraga, M. A. 2006. A Conceptual Framework to Assess Sustainability in Urban Ecological Systems. *International Journal of Sustainable Development and World Ecology* 13 (1):1–15.

O'Leary, A. 2012. "Everybody Inhale." *New York Times.* March 1, 2012.

Organization for Economic Cooperation and Development. 1982. *Product Durability and Product Life Extension.* Paris: OECD.

Organization for Economic Cooperation and Development. 2001. *Aggregated Environmental Indices. Review of Aggregation Methodologies in Use.* Paris: OECD Environment Directorate.

Organization for Economic Cooperation and Development. 2011. *OECD Economic Outlook.* Vol. 2011, issue 2. Paris: OECD Publishing. http://dx.doi.org/10.1787/eco_outlook-v2011-2-en.

Organization for Economic Cooperation and Development. 1993. *OECD Core Set of Indicators for Environmental Performance Reviews.* OECD Environment Monographs No. 83. Paris: OECD Publishing.

Parnell, S., E. Pieterse, D. Simon, A. Simone. 2010. *Urbanization Imperatives for Africa.* Rondebosch, South Africa: African Centre for Cities.

Pataki, D. E., R. J. Alig, A. S. Fung, N. E. Golubiewski, C. A. Kenned, E. G. McPherson, D. J. Nowak, R. V. Pouyat, and P. R. Lankao. 2006. Urban Ecosystems and the North American Carbon Cycle. *Global Change Biology* 12:1–11.

Pennock, J. L., and C. M. Jaeger. 1964. Household Service Life of Durable Goods. *Journal of Home Economics* 56 (1):22–26.

Perlich, C., F. Provost, and J. S. Simonoff. 2004. Tree Induction vs. Logistic Regression: A Learning-Curve Analysis. *Journal of Machine Learning Research* 4:211–255.

Pflüger, M. 2004. A Simple, Analytically Solvable, Chamberlinian Agglomeration Model. *Regional Science and Urban Economics* 34:565–573.

Pickett, S. T. A., M. L. Cadenasso, and J. M. Grove. 2005. Biocomplexity in Coupled Natural-Human Systems: A Multidimensional Framework. *Ecosystems (New York, N.Y.)* 8: 225–232.

Pickett, S. T. A., M. L. Cadenasso, J. M. Grove, C. H. Nilon, R. V. Pouyat, W. C. Zipperer, and R. Costanza. 2001. Urban Ecological Systems: Linking Terrestrial Ecological, Physical, and Socioeconomic Components of Metropolitan Areas. *Annual Review of Ecology and Systematics* 32:127–157.

Plato. [380 B.C.] 2003. The Republic, Books I–V. Trans. Paul Shorey. Cambridge: Harvard University Press.

Ponting, C. 1991. *A Green History of the World: The Environment and the Collapse of Great Civilizations*. New York: Penguin.

Portney, K. E. 2001. *Taking Sustainable Cities Seriously: A Comparative Analysis of Twenty-Three U.S. Cities*. Paper presented at the 2001 Meetings of the American Political Association, American Political Science Association, San Francisco, August 30–ept 2.

Princen, T. 2005. *The Logic of Sufficiency*. Cambridge, MA: MIT Press.

Provost, F., and T. Fawcett. 2001. Robust Classification for Imprecise Environments. *Machine Learning* 42:203–231.

Provost, F., T. Fawcett, and R. Kohavi. 1998. "The Case Against Accuracy Estimation for Comparing Classifiers." In Proceedings of the Fifteenth International Conference on Machine Learning (ICML-98).

Pumain, D. 1998. Urban Research and Complexity. In *The City and Its Sciences*, ed. C. S. Bertuglia, G. Bianchi, and A. Mela, 323–361. Heidelberg: Physica Verlag.

Reader, J. 2004. *Cities*. New York: Atlantic Monthly Press.

Resilience Alliance. 2007. "Urban Resilience. Research Prospectus: A Resilience Alliance Initiative for Transitioning Urban Systems towards Sustainable Futures." Canberra: Commonwealth Scientific and Industrial Research Organisation.

Rosado, L. 2012. "A Standard Model for Urban Metabolism: Accounting Material Flows in Metropolitan Areas." PhD thesis, Instituto Superior Técnico, Technical University of Lisbon.

Rousseau, D. and Y. Chen. 2001. Sustainability Options for China's Residential Building Sector. *Building Research and Information* 29 (4):293–301.

Runnels, C. N. 1995. Environmental Degradation in Ancient Greece. *Scientific American* 272 (3):96–99. doi:10.1038/scientificamerican0395-96.

Saldivar-Sali, A. N. D. 2010. "A Global Typology of Cities: Classification Tree Analysis of Urban Resource Consumption." Master's Thesis, Building Technology Program, Department of Architecture, MIT.

Sankhe, S., I. Vittal, R. Dobbs, A. Mohan, A. Gulati, J. Ablett, et al. 2010. "India's Urban Awakening: Building Inclusive Cities." New York: McKinsey Global Institute.

Satterthwaite, D. 2003. The Millennium Development Goals and Urban Poverty Reduction: Great Expectations and Nonsense Statistics. *Environment and Urbanization* 15 (2): 181–190.

Schaeffer, R. 1996. The embodiment of carbon associated with Brazilian imports and exports. *Energy Conversion and Management* 37:955–960.

Schandl, H., and N. Eisenmenger. 2006. Regional Patterns in Global Resource Extraction. *Journal of Industrial Ecology* 10 (4):133–147.

Sivakumar, A. 2007. "Modelling Transport: A Synthesis of Transport Modeling Methodologies." Imperial College—London and BP-British Petroleum.

Smith, M. 1998. Perspectives on the U.S. Paper Industry and Sustainable Production. *Journal of Industrial Ecology* 1 (3):69–85.

Smith, R. J. 1973. Medium-Term Forecasts Reassessed: IV. Domestic Appliances. *National Institute Economic Review* 64 (1):68–83.

Sousa Pinto, M. J. 2005. *Levantamento Cartográfico de Locais de Pedreiras no Concelho de Lisboa* [Mapping of quarry sites in Lisbon City]. Lisboa: Câmara Municipal de Lisboa—Pelouro do Licenciamento Urbanístico Reabilitação Urbana, Planeamento Urbano e Planeamento Estratégico.

Sterman, J. 2000. *Business Dynamics: Systems Thinking and Modeling for a Complex World.* New York: McGraw-Hill.

Strogatz, S. 2009. "Guest Column: Math and the City." The Wild Side, *New York Times,* March 19, 2009.

Suh, S. 2004. Functions, Commodities and Environmental Impacts in an Ecological Economic Model. *Ecological Economics* 48 (4):451–467.

Sustainable Europe Research Institute. 2008. Global resource extraction 1980 to 2005. Online database. Vienna: SERI. http://www.materialflows.net/mfa/index2.php (accessed 10/2009).

Taylor, Jr., H.A. 1996. *Dimension Stone.* Washington: Publication of the U.S. Geological Survey, Data Series 140, last modification November 17, 2011.

United Nations. 1992. *Report of the United Nations Conference on Environment and Development: Annex I.* New York: United Nations Department of Economic and Social Affairs.

United Nations. 2000. *Integrated Environmental and Economic Accounting: An Operational Manual.* New York: United Nations.

United Nations. 2009. *World Population Prospects: The 2008 Revision.* New York: Department of Economic and Social Affairs, Population Division, United Nations.

United Nations. 2010. *The State of African Cities 2010: Governance, Inequality and Urban Land Markets.* Nairobi, Kenya: UN Habitat, UNEP: November 2010.

United Nations-Habitat. 2008. *State of the World's Cities 2008/2009.* London: Earthscan.

United States Geological Survey. 1996. *Mineral Commodity Summaries: Stone (Dimension).* Washington: Publication of the U.S. Geological Survey, January 1996.

United States Geological Survey. 2011. *Stone (Dimension) Statistics, Compiled by C.A. DiFrancesco (retired) and T.P. Dolley, Data Series 140, last modification November 17, 2011.* Washington: U.S. Geological Survey, U.S. Department of the Interior.

Urban World. 2011. "Toronto Races Ahead with New Green Roof By-Law." In NEWS: North America. *Urban World-UN Habitat* IV (4):43.

van den Bergh, J. C. J. M., and M. A. Janssen. 2004. *Economics of Industrial Ecology.* Toronto: University of Toronto Press.

Vitousek, P. M., H. A. Mooney, J. Lubchenco, and J. M. Melillo. 1997. Human Domination of Earth's Ecosystems. *Science* 277:494–499.

von Bertalanffy, L. 1968. *General System Theory: Foundations, Development, Applications.* New York: George Braziller.

von Wright, G. H. 1989. Science, Reason and Value. Jubilee Lecture of the Royal Swedish Academy of Sciences. Documenta 49. Stockholm: RSAS.

Wackernagel, M., and W. E. Rees. 1996. *Our Ecological Footprint: Reducing Human Impact on the Earth.* Gabriola Island: New Society Publishers.

Waddell, P. 2002. UrbanSim: Modeling Urban Development for Land Use, Transportation and Environmental Planning. *Journal of the American Planning Association* 68 (3): 297–314.

Wegener, M. 1994. Operational Urban Models: State of the Art. *Journal of the American Planning Association* 60 (1):17–29.

Wei, Y. 2005. Planning Chinese Cities: The Limits of Transitional Institutions. *Urban Geography* 26 (3):200–221.

Weisz, H. and H. Schandl. 2008. Material Use across World Regions. *Journal of Industrial Ecology* 12 (5/6): 629–636.

Wernick, I., R. Herman, S. Govind, and J. Ausubel. 1997. Materialization and Dematerialization: Measures and Trends. In *Technological Trajectories and the Human Environment*, eds. J. Ausubel and H. D. Langford, 135–156. Washington: National Academy Press.

Wheeler, S. 2000. Planning for Metropolitan Sustainability. *Journal of Planning Education and Research* 20 (2):133.

Woetzel, J., L. Mendonca, J. Devan, S. Negri, Y. Hu, L. Jordan, X. Li, et al. 2009. *Preparing for China's Urban Billion.* New York: McKinsey Global Institute.

Wolman, A. 1965. The Metabolism of Cities. *Scientific American* 213 (3):179–190.

World Commission on Environment and Development. 1987. *Our Common Future.* New York: Oxford University Press.

World Wildlife Fund. 2011. The Energy Report: 100% Renewable Energy by 2050. WWF.

Xinhua. 2006. "China Mandates Energy Efficiency Standard in Urban Construction." *People's Daily* (Beijing) February 25. http://english.peopledaily.com.cn/200602/25/eng20060225_245935.html.

You, N., and M. Takelman. 2011. "How Chengdu is Building a Bridge to the Urban-Rural Divide." *Urban World-UN Habitat* 4 (4):43.

Zhu, Y. and B. Lin. 2004. Sustainable Housing and Urban Construction in China. *Energy and Buildings* 36 (12):1287–1297.

Zipf, G. K. 1949. *Human Behavior and the Principle of Least Effort.* London: Addison-Wesley.

Green City Plans, Organizations, and Initiatives

Austin, Texas: see various sites related to Austin's Smart Growth Initiative and Transit-Oriented Development: http://www.ci.austin.tx.us/smartgrowth and http://www.ci.austin.tx.us/planning/tod

C40 Cities: Climate Leadership Group. http://live.c40cities.org

CONCERTO Initative, European Commission. http://www.concertoplus.eu/cms/index.php ?option=com_content&view=frontpage&Itemid=113&lang=en

Covenant of Mayors. http://www.eumayors.eu

Energy and Urban Planning in Restructuring Areas. http://www.enpire.eu/index.aspx

European Energy Award. http://www.european-energy-award.org/About-eea-R.58.0.html

European Green Capital Award. http://ec.europa.eu/environment/europeangreencapital/green _cities.html

European Green Cities Network. http://www.europeangreencities.com/about/about.asp

European Smart Cities. http://www.smart-cities.eu

Green Solar Cities. http://www.greensolarcities.com

International Council for Local Environmental Initiatives. www.iclei.org

Portland, Oregon: see various sites related to Portland's initiatives, beginning with http:// www.portlandoregon.gov/bps/28534

Smarter Cities. A project of the Natural Resources Defense Council. http://smartercities.nrdc.org/ rankings/scoring-criteria

Solar Cities, E. C. http://www.isci-cities.org/

SustainLane. http://www.sustainlane.com/us-city-rankings

United States Green Building Council. www.usgbc.org

World Mayor Council on Climate Change. http://www.iclei.org/index.php?id=7192

World Mayors and Local Governments Climate Protection Agreement. http://www .globalclimateagreement.org

Index